WIRING
SIMPLIFIED

by
H. P. Richter

as revised by

W. Creighton Schwan, P.E.

ICBO-IAEI Certified Electrical Inspector
Member of
International Brotherhood of Electrical Workers
International Association of Electrical Inspectors
National Fire Protection Association

33rd Edition

All wiring methods shown in this book
are based on the latest (1981)
National Electrical Code

PARK PUBLISHING, INC.

ST. PAUL, MINNESOTA

PUBLISHED BY

PARK PUBLISHING, INC.

1999 SHEPARD ROAD
ST. PAUL, MINNESOTA 55116

Copyright 1959, 1962, 1965, 1968, 1971, 1974, 1977, 1981
Park Publishing, Inc.
Library of Congress Card No. 33-7980

ISBN: 0-9603294-1-2

Printed in U.S.A.

Printed by The Webb Company, St. Paul, Minnesota

TABLE OF CONTENTS

CHAPTER

PREFACE

This book has been written for people who want to learn how to install electrical wiring, so that the finished job will be both practical and safe. The installation will comply with the National Electrical Code. Then the finished job will be acceptable to electrical inspectors, power suppliers, and others having jurisdiction in the matter.

Electrical wiring cannot be learned by skimming through this or any other book for 15 minutes; neither should the book be considered a "Little Giant Quick Answer" reference where you can find the answer to your question on Page so-and-so. Nevertheless, careful *study* of this book should enable you to wire a house or a farm, so that it will be acceptable to everyone concerned. However, *before doing any wiring, learn how to do the job correctly.*

The author hopes this book will also be of considerable value in *planning* a wiring job, to enable you to write sensible specifications which will lead to your securing maximum usefulness from electric power — now, and 10 years later. Careful planning will avoid later changes which usually cost several times as much as when included in the original plans.

Throughout this book I have emphasized the reasons *why* things are done in a particular way. This will help you to understand not only the exact problems discussed, but will also help you to solve other problems as they arise in actual wiring of all kinds.

The 33rd edition of *Wiring Simplified* has been, and all future editions will be, prepared by W. Creighton Schwan, who is especially well qualified to do so. I am very grateful to him for undertaking this project.

H. P. RICHTER

PREFACE

The ability of H. P. Richter to reduce a complex subject to simple, easy to understand, terms is rare. Matching that ability is improbable, but is a challenge I seriously accept. Revisions in this 33rd edition reflect those made necessary by the many revisions in the 1981 National Electrical Code, and by materials and methods not covered in previous editions. Suggestions for improvement (especially from teachers who use it as a textbook) will be greatly appreciated.

W. CREIGHTON SCHWAN

Chapter 1

STANDARDS, CODES AND SAFETY

Electric power helps us in a thousand ways: it lights our homes, entertains us with radio and TV, operates our kitchen appliances; it runs motors; it takes drudgery out of many farm chores. All this is done with little danger to the user, and so we take electrical power for granted. Yet electric power can cause fires, kill people, and destroy property. It does its work *safely* only because in our homes and in industry it is *under control*. Safety also requires using appliances, lamps and other equipment of dependable quality when installed, and then properly maintained.

The multitudes of different electrical parts and devices in use today are not made exactly the way thousands of manufacturers think they should be made; they are not installed in any fashion that workmen may wish to install them. They are built to entirely definite *safety* standards; they are installed in a manner which is quite uniform throughout all states, a manner which experience has proved practical and safe.

Underwriters Laboratories Inc: All this did not just happen. Reputable merchants and manufacturers will sell only merchandise that is "Listed* by Underwriters Laboratories Inc.", or "Listed by UL" as it is commonly abbreviated. Underwriters Laboratories Inc. is an independent not-for-profit testing laboratory. A manufacturer submits his product to UL; the product is investigated and subjected to performance tests and if it meets the *safety* standards, it is then "Listed by UL." Trained inspectors will periodically visit the factory where the product is made, to audit production controls to determine that the product continues to meet requirements. Additionally, UL regularly tests samples from the factory, or bought in stores. If these samples meet the requirements, the product continues to be listed.

Some items (wire, large switches, fixtures, conduit) have a UL Listing Mark similar to those shown in Fig. 1-1, on each coil or piece. Flexible cord and wire have labels as shown in Fig. 1-2. Devices like toggle switches carry the words "Und. Lab. Inc. List." molded or stamped on each piece. Still other items (receptacles, sockets, outlet boxes, toasters, and similar) have no label but the nameplate, tag or the item itself bears the Listing Mark of Fig. 1-3. Each piece is marked in some way so that the inspector or purchaser can identify the manufacturer by referring to the UL Directory which tabulates manufacturers who have proved that their product meets the UL requirements. The proper UL Listing Mark on the product indicates compliance.

"Listed" means What? An automobile tire intended for use on a passenger car may be entirely suitable and safe for that purpose, but would be totally unsuitable and unsafe if used on a heavy truck. So, too, "Listed by UL" indicates that the merchandise is suitable and safe *if used for the purpose for which it was intended*. It does not mean that the merchandise may be used regardless of circumstances; it may be used only for the
* Many people say "Approved by Underwriters" but that is not correct terminology.

purpose for which it was designed and only under the conditions originally intended. For this reason, an electrical inspector will sometimes turn down listed merchandise; for example, he will refuse to accept listed armored cable if used in a barn because the Code does not permit its use in barns. The inspector will turn down listed lamp cord if used for permanent wiring because it is listed only for use on portable equipment.

Fig. 1-1. Examples of labels applied on merchandise which is listed by Underwriters Laboratories Inc.

Fig. 1-2. Listed cord sets (extension cords and appliance cords) bear one of the labels shown at left.

Important: Listing by UL does not mean that two similar pieces of merchandise are of the same quality. It merely indicates that both pieces meet the UL safety requirements. One may just barely meet those requirements, the other far surpass them. For example, of two brands of toggle switches, one may average 6,000 "ons" and "offs" at full load before breaking down, the other 25,000. Use your own good judgment in making your choice of several listed brands, as you would in selecting other merchandise.

Fig. 1-3. If because of size, shape, material, or surface texture, the product bears only the symbol at left, a complete Listing Mark with the four elements shown at right will be found on the smallest unit container in which the product is packaged.

Codes: Listed electrical parts of high quality, but carelessly or improperly installed, may still be dangerous, both as to shock and fire. For a safe installation, listed devices must be installed as required by the National Electrical Code * (abbreviated NEC). The Code is simply a set of rules which outline the wiring methods that over a period of many years have been found to be safe and sensible. The Code permits installations to be made in several different ways, but all wiring must be done in one of the ways outlined in the Code. A new Code appears every three years. The next edition will be the 1984.

In this book, all references to Code section numbers will be to sections in the 1981 Code, which contains a large number of changes from the 1978 edition. These changes amongst others consist of: (a) Renumbering a section *without* a change in text, so that the only change is in the number of the section; (b) Renumbering a section but *with* changes in the text, sometimes of minor but often of major importance; (c) New section numbers covering material not in the previous edition. Therefore a 1981 section number mentioned in this book may or may not be about the subject covered by the same section number in an earlier or later Code. There are vertical lines in the margins of some pages indicating that the material so marked differs from that in the previous edition.

The National Electrical Code is sometimes supplemented by local Codes or ordinances which are seldom contrary to the National Code, but limit its application. For example, armored cable wiring is one method permitted by the National Code, but sometimes prohibited by local Code.

All methods described in this book are in strict accordance with the latest National Electrical Code, as interpreted by the author. Anyone using this book must recognize that the author does not and can not accept liability from its use. However, much effort has been devoted to making all statements in the book correct, but the final authority on the Code is your local inspector. *This book covers only the wiring of houses and farm buildings; if here and there the statement appears that some particular thing is "always" required by the Code, it means "always so far as the type of wiring described in this book is concerned."*

Study of the Code is necessary and helpful, but the Code alone will not teach you how to wire buildings. Read the following quotation† from the Code: "Purpose: *a.* The purpose of this Code is the practical safeguarding of persons and property from hazards arising from the use of electricity. *b.* This Code contains provisions considered necessary for *safety.* Compliance therewith and proper maintenance will result in an installation essentially free from hazard, but not necessarily efficient, convenient, or adequate for good service or future expansion of electrical use. *c.* This Code is not intended as a design specification nor an instruction manual for untrained persons."

Legal Aspects: Neither the UL requirements nor the Code have the force of law. However, states and cities pass laws which require that only UL listed materials may be used in wiring, and that all wiring must be in accordance with the Code. R.E.A. projects require the same. Usually power suppliers will not furnish power to buildings which

---

* Every student is urged to study the Code. You can get a copy by sending $8.25 to National Fire Protection Assn., Batterymarch Park, Quincy, Massachusetts 02269. Ask for "1981 NEC."

†Excerpted with permission from the National Electrical Code, 1981 edition, copyright 1980, National Fire Protection Association, Boston, MA 02210.

have not been properly wired, and insurance companies may refuse to issue policies on buildings not properly wired. Therefore you have no choice except to follow the law, and in doing so conscientiously you will automatically produce an installation which is safe.

Permits: In many places it is necessary to get a permit from city, county, or state authorities before a wiring job can be started. The fees charged for permits generally are used to pay the expenses of electrical inspectors, whose work leads to safe, properly installed jobs. Power suppliers usually will not furnish power until an inspection certificate has been turned in.

Licenses: Many areas have laws which require that no one may engage in the *business* of electrical wiring, without being licensed. Does that mean that you can not do electrical work *on your own premises* without being licensed? In some localities, the law is so interpreted, in others it is not so interpreted. Consult your power supplier or local authorities.

But remember that if a permit is required, you must get that before proceeding. Before applying for a permit, be sure you understand all problems in connection with your job, so that your wiring will meet national and local Code requirements. If it does not, your inspector must turn down your job until any errors have been corrected.

Safety: Electrical systems in buildings must be installed in such a way that they present the least possible hazard to the occupants or the property. In addition, the work itself must be performed safely so as to protect the installer. Unless you feel confident you can satisfy both of these safety objectives, it would be best that you use this book as a guide to understanding your wiring system, and leave the actual installation to a professional.

Disconnect the Circuit: Some testing can be done only on energized circuits. Be extremely careful when testing that you not contact live parts. Whenever possible, work should be done *only* on circuits which have been de-energized; disconnected from the source of power. If the disconnecting means is out of sight from the work location, you must take positive steps to insure that someone else cannot inadvertently energize the circuit while you are working on it.

Eye Protection: Cut ends of wire, hot solder or flux, and plaster dust have a way of making right for the eyes. Should an arc inadvertently be started, molten metal can be thrown out much too quickly for you to escape it. You have only one pair of eyes. Protect them *at all times* by wearing safety glasses whenever you are doing any electrical work.

Chapter 2

MAKE A "GOOD BUY" IN YOUR ELECTRICAL INSTALLATION

At one time or another you have bought something that you felt was a "good buy"; later you were disappointed, felt that you could have done better. Plan your electrical installation so that you will not be disappointed later, but will still be pleased with it five or fifteen years or more after it is completed. Do not skimp on the original installation. *To add outlets to an installation later usually costs several times as much as including them in the original job.*

In planning your installation, look ahead to the equipment that you are going to be using five or even fifteen years from now. Consider electric cooking, water heating, air conditioning, and even appliances that are not on the market today. Do that, and you will have what is known as "adequate wiring"; you will be making a "good buy" in your wiring job.

If in wiring your buildings, you follow the Code strictly, you will produce a *safe* installation. But it will not necessarily be a convenient, efficient or practical installation. That will come about only if you most carefully plan your job.

Install Large Service Entrance: All the power you use comes into the building through the service entrance wires and related equipment. Start your planning with an adequate service entrance, for the service wires are like a highway: a two-lane highway will not handle four-lane traffic, and neither will small wires carry a large load satisfactorily. This will be discussed in more detail later.

Install Enough Lighting: An adequately wired home is first of all well lighted. The lighting will be from floor and table lamps in some rooms, from permanently installed fixtures in others. Lighting fixtures do not need to be expensive or elaborate, but you must have outlets for fixtures so located that sufficient light will be provided where needed.

Install Lots of Switches: The 1975 Code in Sec. 210-26(a) was the first to require that every habitable room of a house, also hallways, stairways and attached garages, must have some lighting controlled by a wall switch. In kitchens and bathrooms the switch must control permanently-installed lighting fixtures; in other rooms the switch may control one or more receptacle outlets, into which floor or table lamps can be plugged. Each entrance into the house must also have an outdoor light controlled by a switch inside the house. But regardless of Code requirements, an adequately-wired house will have wall switches so located that you can enter by the front door or back, go from basement to attic, without ever being in darkness, and still not leave lights turned on behind you. It does cost a few dollars more to install a wall switch instead of a pull-chain. Entirely aside from the added convenience, consider wall switches as accident insurance against stumbling over something while looking for a pullchain in the dark. To install a pair of 3-way switches so that a light can be turned on or off from two different places (as for example upstairs or downstairs) costs only a few dollars more than the cost of a single switch, and is well worth the difference. In any case, provide convenient switching of lighting, as much energy is wasted by leaving lights on when they are not needed. The energy crisis is here, and will probably get worse before it gets better, so make it a habit — turn off light whenever it is not needed.

Install Lots of Receptacles: An adequately wired house has enough plug-in outlets shown in Fig. 2-2 so that you will never need an extension cord for a lamp, clock or similar equipment. Extension cords are useful as temporary devices, but should never be used to permanently carry power from a receptacle to lamps or other equipment. If there are circumstances where you cannot avoid using them, *never* run a cord under a rug or carpet, for that is inviting a fire. Never run one across open space on a floor, for stepping on it is likely to result in damage and fire, entirely aside from the danger of tripping on it. Under no circumstances ever tack or staple a cord to a wall. Best of all, get along without cords except for *temporary* use.

The Code in Sec. 210-52(a) requires that "in every kitchen, family room, dining room, living room, parlor, library, den, sun room, bedroom, recreation room, or similar rooms of dwelling units, receptacle outlets shall be installed so that no point along the floor line in any wall space is more than six feet, measured horizontally, from an outlet in that space, including any wall space two feet or more in width and the wall space occupied by sliding panels in exterior walls."* This is a minimum, and especially living rooms should have more outlets. A really adequately-wired house will have receptacles so that a floor lamp, for example, can be placed anywhere along the wall without requiring an extension cord.

Wire Size: The Code does not permit a wire smaller than No. 14, protected by a 15-amp. fuse or circuit breaker, in house wiring (except for chimes, thermostats, and other low-voltage wiring). There is a trend towards using No. 12 wire, with 20-amp. protection, and in a few localities it is required as a minimum. The larger wire means brighter lights, less power wasted in heating of wires, fuses that blow less often. The use of No. 12 in place of No. 14 adds very little to the cost of the installation, and will prove a good investment. See also "Small Appliance Circuits" in Chap. 5.

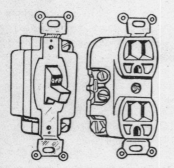

Fig. 2-1. A typical toggle switch. The "plaster ears" on the ends of the strap are a great convenience in mounting the switch.

Fig. 2-2. A typical duplex receptacle. It has two terminal screws on each side. It is rated at 15 amp.

* Reproduced with permission from the National Electrical Code, 1981 edition, copyright 1980, National Fire Protection Association.

Living Room: Suit yourself about whether to have a fixture in the center of the ceiling or on a wall. Some homes depend entirely on floor and table lamps for light. If you do depend entirely on lamps, have some of the receptacles into which they are plugged, controlled by wall switches, so that all the lamps in the room can be turned off at one time, as will be explained later.

No matter how small the room, there should be a minimum of five receptacle outlets: one for radio or TV, one for electric clock, two for lamps. Locate another where it will always be completely accessible for the vacuum cleaner, regardless of the arrangement of the furniture. Many living rooms will be large enough to require more than five receptacles, based on the "6-foot" rule. Regardless of room size, consider the probable arrangement of the furniture: place receptacles so that they will not be hidden behind furniture, difficult to reach. This may require an extra receptacle or two, but will be a good investment.

Dining Room: One ceiling outlet for a fixture should be provided; it must be controlled by a wall switch. Locate this fixture above the center of the dining room table, rather than the center of the room. The "6-foot" rule usually will require not less than four receptacles. Again be careful to locate them so they will not be hidden behind furniture. One should be always accessible from the dining room table, for use with electric percolators, electric knives, and such.

Kitchen: Since the housewife probably spends more time in the kitchen than in any other room, surely the kitchen deserves really adequate wiring. First of all, let it be well lighted. A fixture in the center of the room must be provided for general lighting; it must be controlled by a wall switch. If there are two doors by which the kitchen can be entered, use a pair of 3-way switches so that the light can be controlled from either door. Another light above the sink is necessary, otherwise anybody working at the sink will be standing in his own shadow.

If a fixture is installed at a place where it is possible to touch the fixture and also a water faucet or other grounded surface at the same time, the fixture should be made of porcelain, or if of metal must be grounded (See Chap. 7). In such locations, fixtures should be controlled by a wall switch. These are safety measures. An ungrounded metal fixture or metal pull chain could become energized through a fault in the fixture, and make a serious shock possible.

Install a generous number of receptacles. As will be explained in Chap. 5, the Code requires two special circuits to handle only receptacle outlets for small appliances including refrigeration equipment in kitchen, pantry, family room, dining room, breakfast room; in most houses most of the receptacle outlets on these circuits will be installed in the kitchen. Install one where the refrigerator is to be placed. Locate others 6 to 10 in. above counter level for toaster, iron, mixer and similar appliances; place them so that it will be convenient to use several appliances at one time. The Code requires a minimum of one receptacle for each counter space wider than 12 in. If any space is separated from another by a sink or range top, each must be considered a separate space. A wide space deserves more than one receptacle.

Even if you are not installing an electric range right away, it is wise to install a heavy-duty receptacle for it at the time of the original installation; it will cost much more to do it later. If appliances such as automatic washer, clothes dryer, deep-freeze, and so on are also to be installed in the kitchen, additional special receptacles for these appliances must be installed. See Chapter 14.

Bedrooms: Some people want a ceiling light in each bedroom; others do not. But if you do install a ceiling light, be sure it is controlled by a wall switch near the door. If there is to be no ceiling light, be sure some of the receptacles are controlled by a wall switch. The "6-foot" rule will probably require four receptacles. Again be careful as to their location: one should be permanently accessible for the vacuum cleaner, and another near the bed for reading lamp, heating pad, and so on. Do not overlook lights in closets. Simple single-lamp fixtures are suitable because most closets are so small that it is easy to find the pull chain controlling the light. Install the light on the ceiling of the closet, or on the wall *above* the door. It must be installed so that clothing or other items stored in the closet cannot be closer than 18 in. to the lamp, or not closer than 6 in. to a recessed incandescent fixture with a solid lens or a surface-mounted fluorescent fixture. This is a safety measure required by Code. Many fires can be traced to clothing in closets touching bare lamps of lighting fixtures. The temperature of the glass bulb of a lamp is often higher than 400° F.

Bathroom: A ceiling light controlled by a wall switch is of course essential. One additional light above the mirror over the basin is often installed, but that is not good lighting. For applying makeup or shaving, two lights are needed, one on each side of the mirror. Fluorescent brackets are ideal for the purpose. One GFCI protected receptacle is required near the basin, where it is convenient for an electric razor, etc. Another located elsewhere will be found convenient (See Chap. 7 for discussion of GFCI). But avoid receptacles near the bathtub. People are killed each year while in a tub, by dropping an appliance into the tub. The appliance may not be defective, but is just as dangerous as a defective one would be. A built in electric heater is both safer and more convenient than a portable one.

Miscellaneous: Where there are stairs, locate a fixture so that it will light every step; use 3-way switches so that the light can be turned on and off from either the top or the bottom of the stairs. This is inexpensive insurance against accidents. Every hall deserves a light. If it is a long hall, install 3-way switches so that the light can be controlled from either end. Install a receptacle for vacuum cleaner. Outdoor lights at front and back doors, controlled by switches indoors, are absolutely essential. In the basement install numerous receptacles for washing machine, workshop motors, and similar use.

Lighting Fixtures: The finest light in the world is natural sunlight. At night we use floor or table lamps, or lighting fixtures, to provide light to replace sunlight. Fixtures can be placed into two broad categories: (a) Those which are primarily decorative, and (b) Those which provide good illumination.

If you have good floor and table lamps, to provide plenty of light for reading, sewing and other work that requires good light, there is no reason why fixtures in the same room can not be of the decorative kind; they provide some general lighting of the area, but not enough for exacting work. Many times one fixture will be much better from a lighting standpoint, than another costing several times as much. Let us consider briefly:

what constitutes good lighting?

It is well known that a very large part of the population has defective eyesight, as is evidenced by the number of people wearing glasses. Authorities disagree on the subject, but poor lighting was probably at least part of the cause. Good lighting today is within the reach of all, for good lighting fixtures and lamps are not necessarily expensive, lamps* are cheap, and in many places a 100-watt lamp can be burned for two hours for a penny.

Good lighting requires not only enough light but also requires proper distribution of light throughout the room. A fixture with exposed lamps that you can see from any part of the room does not produce good lighting. You have found that it is easier to read in the shade of a tree than in direct sunlight, even if there is less light in the shade. In direct sunlight the light all comes from one point, the sun; in the shade of a tree it comes from all directions. So also when using fixtures with exposed lamps, the light all comes from one point, causing sharp shadows and glare, which makes reading or sewing or other fine work impossible without eyestrain. Why? Consider how the eye operates. You know that the pupil of the eye changes in size; if there is much light it becomes smaller, to reduce the amount of light entering the eye. If there is not much light, the pupil enlarges to admit all the light possible.

When the light source is very bright, as with exposed lamps, and other parts of the room are naturally much darker, as you look from place to place in the room your eye muscles are overworked trying to adjust to these widely varying conditions. The result is eyestrain. Provide "shade-of-the-tree" well-diffused light, by using fixtures that do not have exposed lamps.

Larger lamps are more efficient, give more light *per watt of power used,* than smaller sizes. Three 60-watt lamps (total 180 watts) give 10% more light than five 40-watt lamps (total 200 watts). One 150-watt lamp gives twice as much light as five 25-watt lamps (125 watts). One 100-watt lamp gives 15% more light than three 40-watt lamps (120 watts). From these figures it should be plain that a fixture that uses one large lamp will in general provide more light than one using several small lamps of equivalent wattage.

One important point to remember in connection with all lighting: your ceilings should be of a light color. A white ceiling will reflect most of the light thrown against it. Ivory is not quite as good for the purpose. A flat finish is better than a glossy finish.

Buy the styles of fixtures that suit your taste and your purse. Always make sure the fixtures you select are electrically good and safe; look for the Underwriters' label on each fixture. Special care should be taken in the selection and placement of fixtures in locations where the seeing task is critical, such as sewing and reading.

"Long-life" Lamps: Ordinary lamps have an average life of 750 to 1000 hours, if used *on circuits of the voltage stamped on the lamps,* the voltage for which they were designed. Some people think they should last longer. Now it is a simple matter to increase the life of a lamp: just burn it at a voltage lower than it was designed for. For example, a lamp designed for 135 volts but burned at 120 volts will last about four times longer; one designed for 140 volts but burned at 120 volts will last about eight times

* Bulbs? Lamps? Lamp bulbs? Different people call them by different names. Technically the glass part is the bulb; the device is a lamp. In this book they will be called lamps.

longer. But there is a catch: the light *per watt* is reduced by 20% to 30%; the cost of the light is increased by 25% to 35%. Sometimes too such "long-life" lamps consume more watts than the number stamped on them. If such a lamp were marked "60 watts" but actually consumed 70 watts, it would *seem* to produce just as much light as a standard 60-watt lamp with 1000-hour life.

There is a place for such lamps, in factories or other places, if they are located where it is difficult and expensive to replace them when burned out. The increased cost of the light is offset by savings in expensive labor required for replacement. But such lamps do not belong in homes. Such lamps properly marked "Extended Service" are available in recognized brands at a price only a little higher than ordinary lamps. But if you see such "long-life" lamps advertised in newspapers or magazines, for home use, by unknown manufacturers at prices three or four times that of ordinary lamps, feel sure they are 135 or 140 volt lamps marked "115 volts" or "120 volts". Don't buy them; they are poor bargains indeed.

Fluorescent Lighting: This form of lighting has many advantages over "incandescent lighting" — ordinary lamps. First of all it produces far more light per watt of power used. Fluorescent lamps last much longer than ordinary lamps.

The life of fluorescent lamps depends mostly on the number of burning hours per start. If fluorescent lights are turned on and left burning continuously, they will last at least 8 to 15 times as long as ordinary lamps. Even if turned on and off frequently, the fluorescent will last 3 to 5 times as long as an ordinary lamp. Moreover the surface brightness of fluorescent lamps is much lower than that of filament lamps, thus providing light that is more diffused, more nearly the "shade-of-a-tree" variety, which is much more comfortable.

Fluorescent lamps used in homes come in lengths from 18 to 60 in., consuming from 15 to 60 watts. The 48-in. 40-watt is the most common. The electric current does not flow through a filament as in ordinary lamps, but rather jumps as an arc from a contact in one end of the lamp, through a gas in the glass tube, to another contact at the other end of the lamp. Now if you had such a fluorescent lamp made of *clear* glass in a dark room, you would see only a slight blue glow when the lamp was turned on. Why? The arc in the lamp provides mostly ultra-violet light which is invisible. So the inside of the lamp is coated with a peculiar whitish powder; when the invisible ultra-violet light strikes this powder it makes the powder "fluoresce" (glow), producing the visible light which we see.

The electrical operation of a fluorescent lamp is a bit complicated. The complete circuit includes the lamp, a ballast and a starter. The most common fluorescent lamp has a filament or "cathode", as it is called, at each end, but these are lighted only for a second or so when the light is first turned on. The ballast is an electrical winding on a steel core so designed that it will do two things: it limits the total power that can flow through it, and also has the peculiarity that when power flowing through it is disconnected, it momentarily delivers a voltage much *higher* than the original voltage flowing through it: for a brief moment it becomes a step-up transformer. The starter is a peculiar kind of switch that opens itself, after power has flowed through it for a moment, and then remains open.

Now see Fig. 2-3, which shows the circuit for the first second or so after the light is turned on. Note that the current flows first through the ballast, then through the filament at one end of the lamp, then through the starter, lastly through the filament at the other end of the lamp, then back to SOURCE. The starter opens itself and after that, there is no circuit through which the current can flow, except through the lamp. As the switch in the starter opens, the ballast delivers a momentary high-voltage kick, and that high voltage is enough to jump the gap inside the lamp; the current then flows through the lamp from end to end, as shown in Fig. 2-4, which shows the circuit after the switch in the starter has opened, and the lamp is in normal operation.

The current flows through the lamp helped by a bit of mercury in the tube, which vaporizes when the lamp goes into operation. The ballast limits the current through the lamp, keeping it at a proper value.

Some kinds of fluorescent lamps are designed to start in a somewhat different way; there are for example the rapid-start and instant-start types. The basic principle of operation is the same in all varieties: a high-voltage kick to start the lamp, with a ballast to limit the current to the proper value.

The diagrams of Figs. 2-3 and 2-4 are for a fixture with only one lamp. The diagrams for fixtures with two or more lamps are of course a good deal more complicated.

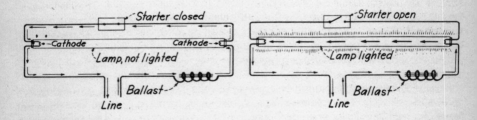

Fig. 2-3. The current has just been turned on. The starter is still closed.

Fig. 2-4. The starter has opened, current flows through the tube; the lamp has been lighted.

Fluorescent lamps are available in a somewhat confusing assortment of different kinds of "white," amongst them deluxe warm white, warm white, white, cool white, and deluxe cool white. The cool white is most commonly used. The cool white produces light most nearly like natural light, tends to emphasize the blue and green color in objects lighted; the warm white produces light more nearly like that produced by ordinary filament lamps, and tends to emphasize the red and brown in objects lighted. The deluxe varieties emphasize these qualities still more, but are a bit less efficient, producing a little less light per watt of power used.

Chapter 3

MEASUREMENT OF ELECTRICITY

Water is measured in gallons, wheat in bushels, meat in pounds. Electric power cannot be poured into a measure or weighed on scales, but rather is something which must be considered as always in motion. The problem is to measure how much flows past any point, at any given moment, or in total over a period of time.

Amperes: When we want to measure a quantity of water, we talk of "gallons"; the absolute measure of a quantity of electric power is the "coulomb." When we want to talk of water in motion, we talk of "gallons per second"; in the case of electric power we talk of "coulombs per second." Yet few people have ever heard the word "coulomb." That is because we use the simpler term "ampere," because when current flows at the rate of "one coulomb per second" we say simply "one ampere." Do not say "amperes per second," say just "amperes."

For those who are electronically inclined, let it be said that an electric current is merely the flow of electrons past a given point. One ampere is the movement of 6.28 billion billion (6 280 000 000 000 000 000) electrons per second.

Volts: Water and air and other substances can be put under pressure; it is common to speak of such pressure in terms of pounds per square inch. Electric power is also under pressure, and pressure is measured in volts. Any ordinary dry cell or flashlight cell when new develops a pressure of about 1½ volts. One cell of an automobile battery develops 2 volts, six cells together develop 12 volts. Most house and farm wiring is at 120* and 240 volts. The voltage at which power is transmitted over high-voltage lines varies from 2400 volts for short distances to 500 000 and more volts for long distances.

Watts and Kilowatts: Amperes alone or volts alone do not tell us the actual amount of power flowing in a wire, any more than "gallons per minute" tell us much about the work involved in pumping water unless we also know the pressure, or the depth of the well from which the water is being pumped, which is almost the same as pressure. Both amperes and volts must be considered. The two together tell us how much power is flowing, for *volts* × *amperes* = *watts.*†

Since watts = volts × amperes, any wattage may consist of either a low voltage and a high amperage, or a higher voltage and a lower amperage. A lamp which draws 5 amp. from a 12-volt battery consumers 5 × 12 or 60 watts; another lamp which draws ½ amp. from a 120-volt line consumers ½ × 120 or 60 watts. The voltage and amperage differ widely, but the actual watts or power consumed by the two lamps is the same.

* 110- and 220-volt current is seldom seen today. The Code uses 115- and 230-volts for calculations, but in this book we use the actual 120- and 240-volts used in most locations.

† The formula is always correct with direct current, but with alternating current is correct only part of the time. It is correct with lamps, ranges, toasters and similar heating appliances. In commonly used equipment it is wrong only with motors, and devices with transformers (radio, TV) or with ballasts (fluorescent lamps), where the watts are somewhat less than volts × amperes.

Watts measure power just as horsepower does; as a matter of fact, 746 watts is always equal to 1 hp. A motor that delivers 1 hp. delivers 746 watts, and could just as well be called a 746-watt motor (it uses more than 746 watts because some power is wasted as heat and it also takes some power to run the motor even when it is not delivering power). A lamp that uses 746 watts could just as well be called a 1 hp. lamp.

A watt is a very small amount of power; when speaking of large amounts of power, it is simpler to speak of kilowatts (the Greek word *kilo* means *thousand*). One kilowatt (abbreviated kW) is 1000 watts.

You must speak of watts, not watts *per hour* or kilowatts *per hour*, just as you say that the engine in your automobile delivers 200 hp., not 200 hp. *per hour.*

Watthours and Kilowatthours: Watts and kilowatts measure the rate at which power is being used at *any given moment.* Watthours and kilowatthours measure the total amount of power that has been used during any specified interval of time. One watt used for 1 hour is 1 watthour. Multiplying the watts used by the number of hours gives the watthours. A 60-watt lamp used for 6 hours consumes 60 × 6 or 360 watthours. A 1000-watt iron used for 2 hours consumes 1000 × 2 or 2000 watthours (2 kilowatthours).

A watthour is a very small amount of power so it is more common to speak of kilowatthours; a kilowatthour is 1000 watthours. The iron mentioned in the previous paragraph, consuming 2000 watthours in 2 hours, consumes 2 kilowatthours, abbreviated kWh. Electric power is measured and paid for by the kilowatthour.

One kilowatthour will operate the average iron for an hour, a washing machine for three hours. It will operate a 1-hp. motor for about an hour, or pump about 1000 gallons of water. It will operate the average radio about 15 hours, a 50-watt lamp 20 hours. It will operate an electric clock for about a month. One kilowatthour will do all this, and at average rates, costs about six cents.

Fig. 3-1. The most modern kilowatt-hour meter has dials like the mileage indicator on the speedometer of an automobile, and is read in the same way.

KILOWATT HOURS
3 3 5 0

Reading Your Meter: Many meters being installed at the present time show the total kilowatthours on a register just like the "odometer" or mileage recorder of the speedometer of an automobile; see Fig. 3-1. The total is easily read.

Reading an old style meter is a little more difficult, but you can learn to do it in a few minutes. Fig. 3-2 shows the register or dials of the meter. Two of the pointers move in one direction, the others in the opposite direction. Assume it is your meter at the beginning of the month. Simply write down, from left to right, the number that each pointer *has passed*, as on a clock. The total of the meter in Fig. 3-2 is 1642 kWh.

The next picture, Fig. 3-3, shows the same meter a month later. One of the pointers points directly to the 7. Before writing down 7, look at the pointer on the dial to the right; it

has not quite reached the "0." Therefore the pointer which seems to point directly to
has not actually reached the 7, so write down 6 instead, making the total reading 226
kWh. (If the last pointer were just past the "0," the total would be 2270). The difference
between the two readings, or 627, represents the number of kilowatthours of electricity
used during the month.

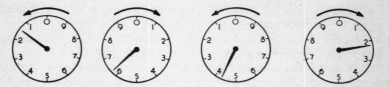

Fig. 3-2. In reading the ordinary meter, write down the number the pointer has passed.

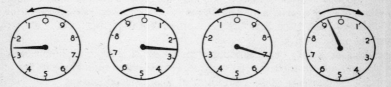

Fig. 3-3. The meter shown in Fig. 3-2, but now a month later.

Power Rates: The rate charged for domestic electric power averages about 6¢ per
kilowatthour in the United States, but varies greatly. Before the energy crisis, rate
schedules were stepped so that the more power you used the lower the average cost
per kWh. Now, in order to encourage conservation of energy, most rates are inverted
the more power you use, the higher the average cost per kWh. A typical rate structure
follows:

First 100 kWh used per month . . . . 3.00¢ per kWh
All over 100 kWh used per month. 6.60¢ per kWh

Your bill for the 627 kWh is figured this way:

100 kWh at 3.00¢. . . . . $ 3.00
527 kWh at 6.60¢ 34.78

627 kWh   Total $37.78
Average per kWh . . . . 6.02¢

Considering the rate of inflation and rising fuel costs, these rates may not be typical at
the time you are reading this.

Operating Cost per Hour: To find out how much it costs to operate any electrical
equipment for one hour, simply multiply the watts the device consumes by the rate in
cents per kWh and point off five places; this gives the cost in dollars per hour. For a
60-watt lamp at 6¢ per kWh, the figures are 60 × 6 = 360; pointing off five places
makes 0.0036 or about ⅓ cents per hour. For any 600-watt appliance at 6¢ per kWh, the
figures are 600 × 6 = 3600; pointing off five decimals makes 0.036, slightly over 3½¢
per hour.

To determine how long any device may be used to consume one kWh, simply divide 1000 by the wattage of the device. For example, a 40-watt lamp may be used 1000 ÷ 40 or 25 hours. A 600-watt appliance may be used 1000 ÷ 600 or 1.666 hours or 1 hour 40 min. An electric clock uses about 2 watts and will run 500 hours while consuming 1 kWh.

Watts Consumed: The following table will aid you in estimating the power required, or the operating cost, for various appliances. The figures are only approximate, and new energy-saving appliances are appearing all the time.

FOOD PREPARATION	Watts		LAUNDRY	Watts
Blender	500 to 1000		Dryer	4000 to 6000
Coffee Maker	500 to 1000		Iron, hand (steam or dry)	600 to 1200
Dishwasher	1000 to 1500		Washing Machine	300 to 550
Frying Pan	1000 to 1200		Washing Machine, automatic	500 to 800
Hot Plate, per burner	600 to 1000		Water Heater	2000 to 5000
Knife	100		**PERSONAL CARE**	
Microwave Oven	1000 to 1500		Hair Dryer	350 to 1200
Mixer	120 to 250		Heating Pad	50 to 75
Oven, separate	4000 to 5000		Shaver	8 to 12
Range	8000 to 14000		**ENTERTAINMENT**	
Range Top, separate	4000 to 8000		Projector, slide or movie	300 to 500
Roaster	1200 to 1650		Radio (tube type)	40 to 150
Rotisserie (broiler)	1200 to 1650		Stereo (solid state)	30 to 100
Toaster	500 to 1200		TV, Black & White (tube type)	150 to 325
Waste Disposer	500 to 900		TV, Black & White (solid state)	50 to 100
FOOD STORAGE			TV, Color (tube type)	300 to 450
Freezer, household	300 to 500		TV, Color (solid state)	150 to 250
Refrigerator, household	150 to 300		**MISCELLANEOUS**	
Refrigerator, frostless	400 to 600		Clock	2 to 3
ENVIRONMENTAL COMFORT			Lamps, Fluorescent	15 to 60
Air Conditioner, Central	2500 to 6000		Lamps, Incandescent	10 upward
Air Conditioner, Room	800 to 2500		Sewing Machine	60 to 90
Blanket	150 to 200		Vacuum Cleaner	250 to 1200
Fan, portable	50 to 200		**MOTORS**	
Heat Lamp (infra-red)	250 to 500		¼-hp.	300 to 400
Heater, portable	1000 to 1500		½-hp.	450 to 600
Heater, wall-mounted	1000 to 4500		Over ½-hp., per hp.	950 to 1000

DC and AC: On a battery, one terminal is always + (positive) and the other − (negative). Any current where each wire is *always* of the same polarity, either positive or negative, is known as direct current or dc. Current from a battery is always dc.

The current used in practically all city homes, and in most "high lines" in the country, is known as alternating current or ac because each wire changes or alternates continually from positive to negative to positive to negative and so on. The change from positive to negative and back again to positive is known as a "cycle." This takes place 60 times every second, and such current is then known as 60-hertz or 60-Hz* current. Sixty times every second each wire is positive, and 60 times every second it is negative, and 120 times every second there is no voltage at all on the wire. The voltage is never constant but is always gradually changing from zero to a maximum of usually about 170 volts, but averaging 120 volts, and such current is then known as 120-volt current. (See fig. 3-4).

What used to be known as 60-cycle current is now called 60-hertz or 60-Hz. In 60-Hz current there are 60 complete cycles every second. The term "hertz" is used instead of "cycles per second"; it is named after the German scientist Hertz who discovered the cyclical nature of electrical waves.

You may well ask the question: if there is no current in the wire 120 times every second
why do not lights flicker? They do not flicker because the filament in the lamp does no
cool off fast enough. Very small lamps used on 25-Hz current (where there is no voltag
on the wire 50 times every second) do have an annoying flicker.

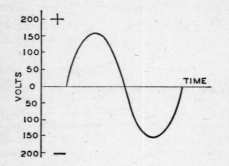

Fig. 3-4. One cycle of 120-v. alternating current. If it is 60-Hz current
all the changes shown take place in
1/60th second.

Single- and 3-Phase Current: The current described in the previous paragraph is
single-phase current. Remember that if 120-volt 60-Hz current flows in a pair of wires
120 times every second the wires are dead, there being no voltage at all; 120 times
every second, the voltage is about 170 volts; at all other times it is somewhere in
between, but averaging 120 volts. It is therefore anything but steady; the voltage is
always changing. The changes however are so rapid that for most purposes it can be
considered a steady 120-volt current.

Imagine now three separate electric generators all on a single shaft, so arranged tha
the voltage reaches its maximum at different times in each of the three generators; firs
in one, then in the second, then in the third, then again in the first, and so on. Run a pai
of wires from each of the three generators. The three generators together then are said
to deliver 3-phase power (although the power from any one generator is stil
single-phase). In actual practice, the three generators become a single generator with
three windings; the three pairs of wires become three wires.

Three-phase power is common in cities in factories and similar establishments where
there are many motors. It is seldom found on farms, for to provide it requires 3-wire
instead of 2-wire high voltage transmission lines; each farm with 3-phase powe
requires three transformers instead of one. Do not be misled into thinking that because
there are three wires in the service entrance, the result must be 3-phase power. On
farms and in homes, almost always three wires mean single-phase 3-wire 120/240-vol
service. (Three-phase power may be supplied with either three or four wires from the
transformers.) If however, you are fortunate enough to have 3-phase power available
by all means use 3-phase motors for greater efficiency. If you do have a 240-vol
3-phase supply, consult the serving agency to determine whether you also have
120/240 volt single phase power for lighting and small appliances without installing you
own transformer.

Chapter 4

WIRE SIZES AND TYPES

Electric power flows through wires. It flows much more easily in some materials than others; copper is the best material for ordinary purposes. If iron wire were used, it would have to be about 10 times as big in cross-sectional area as copper.

All references in this book are to *copper* wire, except the discussion of aluminum later in this chapter.

Copper wire sizes are indicated by number. No. 14 is the most ordinary for house wiring; it is not quite as big as the lead in an ordinary pencil. Nos. 12, 10, 8, and so on are larger than No. 14; Nos. 16, 18, 20 and so on are progressively smaller. No. 14 is the smallest size permitted for ordinary house wiring, and No. 1 is the heaviest usually used in residential and farm wiring. Still heavier sizes are Nos. 1/0, 2/0, 3/0 and 4/0, the 4/0 being almost half an inch in diameter. No. 16 and No. 18 are used mostly in flexible cords and the still finer sizes are used mostly in the manufacture of electrical equipment such as motors. No. 18 is also commonly used in wiring doorbells, chimes, thermostats and similar devices all of which operate at less than 30 volts.

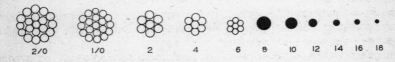

Fig. 4-1. Actual diameters of different sizes of copper wire, without the insulation.

Why Wire Size is Important: Wire of the correct size must be used, this being important for two reasons: ampacity and voltage drop. Ampacity is the safe carrying capacity of a wire, in amperes.

When current flows through wire, it creates a certain amount of heat. The greater the amperes flowing, the greater the heat. (Doubling the amperes without changing the wire size increases the amount of heat *four* times.) This heat is entirely wasted; therefore to avoid wasted power, we must use a wire size which limits the waste to a reasonable figure. Moreover, if the amperage is allowed to become too great, the wire may become so hot that it will damage the insulation or even cause a fire. The Code is not concerned with wasted power, but is concerned with safety; therefore it sets the *ampacity,* or *maximum amperage* that various sizes and types of wires are allowed to carry.

Wire ampacities are shown in the table on the next page. The various types listed in the heading for each column will be explained later in this chapter. Use Columns *A* or *B* for wires in conduit, cable or buried directly in the ground. Use Columns *C, D* or *E* for individual wires in free air, for example, overhead wires from yard pole to building on a farm. If you use Type UF underground cable, use Col. *A*; if Type USE, use Col. *B*. However, when using Type SE cable, sometimes you will find the type of the wire used inside the cable (RH, RHH, RHW, or XHHW) marked on the outside of the cable. When so marked, the temperature rating of the cable is that of the individual conductors. When the conductor type is not marked on the outside, use Col. *B*.

AMPACITY OF COPPER WIRES

Wire Size	In conduit, cable, or buried directly in the earth		Single conductors in free air		
	Types T, TW	Types RH, RHW, THW	Types T, TW	Types RH, RHW, THW	Weather-proof
	A	B	C	D	E
14	15*	15*	20*	20*	30
12	20*	20*	25*	25*	40
10	30	30*	40	40*	55
8	40	50	60	70	70
6	55	65	80	95	100
4	70	85	105	125	130
2	95	115	140	170	175
1/0	125	150	195	230	235
2/0	145	175	225	265	275
3/0	165	200	260	310	320

To find the ampacity of other sizes, conductor metals, or insulations, consult Tables 310-16 to 310-19 in your copy of the Code.

* In the Code the ampacities of these wires are shown higher than this, but a footnote limits their load current rating and overcurrent protection to the figures given here.

If forcing too many amperes through a wire caused only a certain amount of wasted power, we might look upon it as a mere nuisance and loss. However, it also causes *voltage drop*. Actual voltage is lost in the wire so that the voltage across two wires is lower at the end than at the starting point. For example, if you connect two voltmeters into a circuit, as in Fig. 4-2, one at the main switch, one across a 1 hp. motor at a distance, you will find that the voltage at the motor is lower than at the main switch. The meter across the main switch may read 120 volts. If No. 14 wire is used to the motor, the voltage across the motor terminals will be about 119 volts if the motor is 10 ft. away, but only about 112 volts if 100 ft. away. The difference is lost in the wire and is known as voltage drop. Voltage drop is wasted power, but there is one other very important consideration: appliances work very inefficiently on voltages lower than the voltage for which they were designed. At 90% of rated voltage, a motor produces only 81% of normal power; a lamp produces only 70% of its normal light.

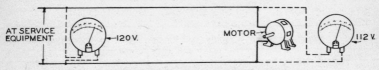

Fig. 4-2. This illustrates "voltage drop." The voltage at the motor is lower than the voltage at the starting point.

In the example above, using 200 ft. of No. 14 wire, the drop was from 120 to 112 volts, or 8 volts, about 7%. If No. 12 wire had been used, the drop would have been reduced about 60%, to 3.2 volts, only about 2½% of the starting voltage. The larger the wire, the less the voltage drop.

Voltage drop can't be reduced to zero, but can be kept at a practical level by using wire of sufficient size. A drop of 2% is considered entirely acceptable. If the starting point is 120 volts, 2% is 2.4 volts, so that the actual voltage at the point where the power is consumed is 117.6 volts. If the starting point is 240 volts, the voltage at the point of consumption is 235.2 volts. The apparent saving in initial cost by using undersize wire is soon offset by the cost of power wasted in the wires, and by the reduction in efficiency of lamps, motors, and so on.

Selecting Right Size Wire: The Code permits nothing smaller than No. 14 for ordinary wiring. It is better to consider No. 12 the smallest, this being required in a few places by local ordinance. If you need wire heavier than the minimum permitted, it is a fairly complicated matter to *figure* the right size, but a simple matter to look it up in tables.

First determine the amperage to be carried by the wire. To help in arriving at the correct figure if motors are involved, use the following table:

Motor	120 v.	240 v.	Motor	120 v.	240 v.
¼ hp.	6 amp.	3 amp.	1½ hp.	20 amp.	10 amp.
⅓ hp.	7 amp.	3½ amp.	2 hp.	24 amp.	12 amp.
½ hp.	10 amp.	5 amp.	3 hp.	34 amp.	17 amp.
¾ hp.	14 amp.	7 amp.	5 hp.	56 amp.	28 amp.
1 hp.	16 amp.	8 amp.			

On the next page appear two tables, one for 120 volts, the other for 240 volts; use the one that corresponds to the voltage of the circuit in question. All distances shown are *one-way* distances; to operate a device 300 ft. away requires 600 ft. of wire, but look for the figure 300 in the table. The distances under each wire size are the distances that size wire will carry the different amperages (or wattages) in the left-hand columns, with the customary 2% voltage drop. For example, in the 120-volt table, to determine how far No. 8 wire will carry 20 amp., follow the 20-amp. line until you come to the No. 8 column, and there is the answer; 90 ft. If a distance is marked with an asterisk (*) it indicates that

Type T or TW wire in conduit or cable, or a cable buried directly in the ground, may not be used, because the amperage in the left-hand column is greater than the ampacity of Type T or TW. Select the proper type of wire from the table on the preceding page. In many cases, only weatherproof wire will have the required ampacity.

TABLE OF WIRE SIZES FOR 120 VOLTS SINGLE-PHASE

Am-peres	Watts at 120 volts	No. 14	No. 12	No. 10	No. 8	No. 6	No. 4	No. 2	No. 1/0	No. 2/0	No. 3/0
5	600	90	140	225	360	570	910				
10	1200	45	70	115	180	285	455	725			
15	1800	30	45	70	120	190	300	480	765	960	
20	2400	20*	35	55	90	145	225	360	575	725	915
25	3000	18*	28*	45	70	115	180	290	460	580	730
30	3600	15*	24*	35	60	95	150	240	385	485	610
40	4800			28*	45	70	115	180	290	360	455
50	6000			23*	36*	55	90	145	230	290	365

TABLE OF WIRE SIZES FOR 240 VOLTS SINGLE-PHASE

Am-peres	Watts at 240 volts	No. 14	No. 12	No. 10	No. 8	No. 6	No. 4	No. 2	No. 1/0	No. 2/0	No. 3/0
5	1200	180	285	455	720	1145					
10	2400	90	140	225	360	570	910	1445			
15	3600	60	95	150	240	380	610	970	1530		
20	4800	45*	70	115	180	285	455	725	1150	1450	
25	6000	35*	55*	90	140	230	365	580	920	1160	1460
30	7200	30*	48*	75	120	190	300	480	770	970	1220
40	9600		36*	56*	90	140	230	360	575	725	915
50	12 000			45*	70*	115	185	285	460	580	725
60	14 400				60*	95*	150	240	385	485	610
70	16 800				50*	80*	130	205	330	410	520
80	19 200					70*	115*	180	285	360	460
90	21 000					60*	100*	160	250	320	405
100	24 000					55*	90*	145*	230	290	365
125	30 000						75*	120*	190	240	300
150	36 000							95*	150*	195*	245
200	48 000							70*	115*	145*	185*

In both tables above, figures represent ONE-WAY distances, not the length of the wire back and forth.

 * For these distances, Type T or TW wires in conduit or cable may not be used because they do not have enough ampacity. For all distances marked with the asterisk, select a type of wire with sufficient ampacity (depending on whether in conduit or cable, or in free air) from the table on page 22.

Each figure indicates the maximum distance in feet each size wire will carry the amperage in the left column, with 2% voltage drop. If you wish to permit 4% drop, double the distances shown. If you wish to permit 5% drop, multiply all distances by 2½.

Compare the 120-volt and the 240-volt tables. You will see that at 240 volts, any given size of wire will carry the same *amperage* twice as far as at 120 volts, with the same percentage of voltage drop; it will carry the same number of *watts* four times as far.

Overhead Wires: When wires are run outdoors overhead, they must of course be big enough to carry the amperage involved, without excessive voltage drop. They must also be big enough to support their own weight. The Code in Sec. 225-6(a) requires a minimum of No. 10 for spans not over 50 ft. and No. 8 for larger distances. For distances over 150 ft. you will be wise to use an extra pole. In northern areas the wires often must support a heavy ice load; it is then often wise to use a size larger than electrically required. If the wire is installed on a hot summer day, leave considerable slack, otherwise the contraction of the wires in winter may pull the insulators off your buildings.

Weatherproof Wire: This kind of wire has a covering of neoprene or impregnated cotton over the conductor, which is *not* recognized as insulation, and may not be used for ordinary wiring; it may be used only overhead outdoors. Its ampacity is as shown in Col. *E* in the Table on page 22. Size for size the ampacity of weatherproof wire is higher than other wires, and it may last longer under exposure to the weather, but remember, the smaller the wire, the greater the voltage drop.

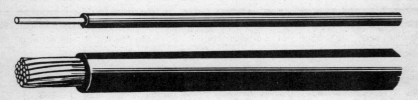

Fig. 4-3. Type T is the most common wire in use. It may be used only in dry locations. If it is No. 6 or larger, it must be stranded; in some cases No. 8 must be stranded.

Wires for Interior Wiring: Several kinds of wire will be described. All are available in various colors, the purpose of which will be described later. (No. 4 and larger however are usually available only in black. Weatherproof wire in all sizes is always black.) If the wire is No. 10 or smaller, it is solid; the copper conductor is a solid strand. If it is No. 8 it may be solid if it is in the form of cable, or is not to be drawn into conduit after installation. But if it is to be drawn into conduit, it must be stranded: a number of smaller wires grouped together to make one larger, more flexible wire, as shown in Fig. 4-1. If the wire is No. 6 or larger, it must always be stranded (except weatherproof).

In use, wires are pulled into pipe called conduit, or used in the form of multi-conductor cables. In addition to the wires described in this book, many other types are available, but not usually used in residential and farm wiring. You will find them listed in your copy of the Code.

Type T Wire: The most commonly used wire is what the Code calls Type T. The conductor is covered by a single layer of plastic compound, the thickness of which depends on the size of the wire, and which strips off easily and cleanly. The wire is relatively small in overall diameter, is clean and easy to handle. See Fig. 4-3. Often

Type TW is used; it is the same as Type T except moisture-resistant, and suitable for use in wet locations. Type THW is both heat- and moisture-resistant. Neither type may be buried directly in the ground.

Rubber-Covered Wires: This kind of wire has a conductor insulated with rubber, either natural or synthetic, followed by an outer fabric braid that is saturated in moisture- and fire-retardant compounds. Except for the black, the color is painted on. A final coat of wax is added for cleanliness. See Fig. 4-4. Once the most popular of all kinds of wire, it is used comparatively little today. Three types however are still available: Types RH and RHH which rarely if ever are used in residential and farm wiring, and Type RHW which is frequently used in the larger sizes. In sizes No. 8 and heavier, the Code assigns a slightly higher ampacity to it, than to Types T and TW. Type RHW may be used in dry or wet locations, but may not be buried directly in the ground.

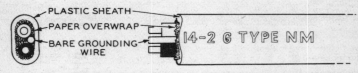

Fig. 4-4. Rubber-covered wire has rubber instead of plastic insulation. It has a fabric overbraid.

Cables: Wires are often assembled into cables such as nonmetallic sheathed cable or armored cable. These will be described in detail later. When a cable contains two No. 14 wires, it is known as 14-2 (fourteen-two) cable; if it has three No. 12, it is called 12-3, and so on. If a cable has for example two No. 14 insulated wires and a bare uninsulated grounding wire, it is known as "14-2 with ground." If a cable has two insulated wires, one is always white, one black; if it contains three, the third is red.

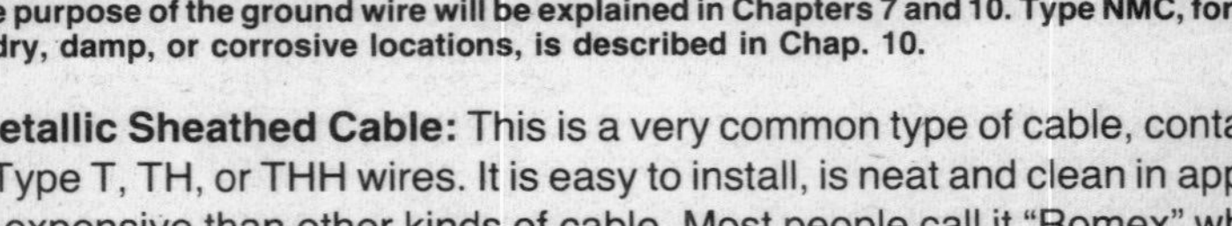

PLASTIC SHEATH
PAPER OVERWRAP
BARE GROUNDING WIRE
14-2 G TYPE NM

Fig. 4-5. Nonmetallic sheathed cable consists of two or three individual wires, assembled into a cable. Type NM 2-wire with ground is shown, and may be used only in dry locations. The purpose of the ground wire will be explained in Chapters 7 and 10. Type NMC, for use in dry, damp, or corrosive locations, is described in Chap. 10.

Nonmetallic Sheathed Cable: This is a very common type of cable, containing two or three Type T, TH, or THH wires. It is easy to install, is neat and clean in appearance, and less expensive than other kinds of cable. Most people call it "Romex" which is the trade name of one particular manufacturer. There are two kinds.

The first is what the Code calls Type NM. As shown in Fig. 4-5, the individually-insulated wires are enclosed in an overall plastic jacket.* Some manufacturers put paper wraps on the individual wires, or over the assembly. Empty spaces between wires are sometimes filled with jute or similar cord. This type may be used only in normally dry locations, never for example in barns on farms.

The second type is what the Code calls Type NMC and is especially designed for

* In older constructions, occasionally still seen, the outer jacket was a braided fabric. In any case, the outer jacket must be moisture-resistant and flame-retardant.

damp or corrosive locations, such as barns and similar locations. It may also be used in ordinary dry locations, of course. It will be described in Chap. 10.

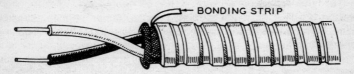

Fig. 4-6. Armored Cable consists of two or three individual wires, assembled into a cable, protected by flexible steel armor. It may be used only in dry locations.

Armored Cable: This type of cable is what the Code calls Type AC, and which most people call "BX" although again that is the trademark of one particular manufacturer. The individual wires are Type T. There is an overwrap of tough paper between the wires and the spiral steel armor. It may be used only in permanently dry locations. The cable is shown in Fig. 4-6.

Underground Wires: See Chapter 17.

Flexible Cords: Flexible cords are used to connect lamps, appliances and other devices to outlets. Each wire consists of many strands of fine wire for flexibility. Over the wire is a wrapping of cotton to prevent the insulation from sticking to the copper. There are many kinds of flexible cords with varying kinds and thicknesses of insulation, depending on the purpose of the cord. The more common kinds will be described here.

Underwriters' Type SPT is the most common cord used for radios, floor lamps and similar devices. As shown in Fig. 4-7, it consists of copper wires imbedded directly in plastic insulation. It is tough, durable and available in various colors. The same kind of cord with a rubber-like insulation is known as Type SP. These cords are commonly available only in No. 18 and No. 16.

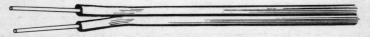

Fig. 4-7. In Type SPT, the wires are imbedded in plastic. The cord is durable, attractive.

Fig. 4-8. Types S and SJ cords are designed for severest use.

A really rough and knock-about cord is Underwriters' Type S, shown in Fig. 4-8. Each wire is rubber- or plastic-insulated: the two wires are then bundled into a round assembly, the empty spaces being filled with jute or paper twine to make the assembly round. Over all comes a layer of tough, high grade rubber. Type SJ is similar except that the outer layer of rubber is thinner. Very similar but more flexible cords used only on vacuum cleaners are called Type SV or SVT.

If the outer jacket is made of neoprene instead of rubber, the cord becomes oil-resistant and the Code designations become SO and SJO, instead of S and SJ.

For irons, toasters and similar appliances delivering considerable heat, "heater cord" is used. Fig. 4-9 shows Code Type HPD, which has a layer of asbestos over each rubber- or thermoplastic-insulated wire, or has all neoprene insulation, plus an outer braid of cotton. Code Type HPN is more popular; it has neither asbestos nor an overbraid; the wires are imbedded in neoprene and its appearance is quite similar to that of Type SPT in Fig. 4-7.

Fig. 4-9. Type HPD cord is used on flatirons, toasters, etc.

Aluminum Wire: This is available in two types: aluminum and copper-clad aluminum (aluminum with a thin sheath of copper on the outside). Aluminum is a relatively poor conductor; a larger size must be used than when using copper. A rule-of-thumb is to use aluminum two sizes heavier than copper: No. 12 aluminum instead of No. 14 copper; No. 4 aluminum instead of No. 6 copper, etc.

When aluminum wire was first used, it was connected to ordinary terminals that were quite suitable for copper, but it soon became evident that they were not suitable for aluminum; the connections heated badly and led to loose connections, excessive heating, and even fires. The Underwriters then required redesigned terminals, marked CU–AL*, which were considered suitable for either copper or aluminum, but it later developed that in the 15- and 20-amp. ratings, they were still not suitable for aluminum. The terminals were further redesigned and since 1971, only devices marked CO/ALR* are acceptable. Note that terminals rated *higher than* 20 amp. were not changed so those marked CU–AL are still acceptable, and *must* be used with aluminum wire.

If your existing installation was made using aluminum, you would be wise indeed to inspect all your receptacles and switches, and if they are not marked CO/ALR, replace them all with devices that are so marked.

Do note these points: (a) If the aluminum wire is copper-clad, any kind of terminal may be used; (b) Push-in terminals shown in Fig. 4-16 may be used with copper, copper-clad aluminum, but *not* with ordinary aluminum.

Removing Insulation: Before wires can be connected to a device or spliced to another piece of wire, the insulation must be removed. There are several ways of doing this. A special tool known as a wire stripper is the most convenient way. Another way is to use a pair of side-cutting pliers. Insert the wire *behind* the hinge of the blades, nearest the hinge, and mash the insulation, softening it, from the point where it is to be removed, to the end. Then place the jaws of the pliers at the point where the removal is to begin, squeeze hard enough to cut into the insulation but *not* touching the conductor, and the mashed insulation can be pulled off easily.

Alternately, you can use a knife. Do not cut the insulation off sharply as shown at A of Fig. 4-10, for there is too much danger of nicking the conductor, leading to later breaks. Hold your knife to produce an angle as shown in B of Fig. 4-10.

---

* The marks CU–AL or CO/ALR are stamped into the mounting yokes of switches, receptacles and similar devices, so as to remain visible without removal from the boxes in which they are installed. On larger equipment the marks are located so as to remain visible after installation.

Always make sure that the stripped end of wire is absolutely clean. Rubber-covered wires have a tinned conductor to make it easy to strip off the insulation. Plastic-insulated wires are not tinned because that kind of insulation strips off cleanly.

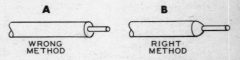

Fig. 4-10. Wrong and right methods of removing insulation from wires.

Terminals: Terminals to which wires are to be connected are of two kinds. One kind consists of a terminal screw in a metal part with upturned lugs to keep the wire from slipping out from under the screw, shown in Fig. 4-11. It may be used with wires No. 10 and smaller, but it is very difficult to make a good connection if the wire is No. 10 and *stranded.* The other kind used mostly for wires larger than No. 10 is shown in Fig. 4-12; the wires are inserted into the terminal, and the screw then tightened. Let's talk first of the kind for smaller wires.

The correct method is shown in Fig. 4-13. Wrap the wire at least two thirds of the way around the screw, preferably three-quarters of the way, in a clockwise direction so that tightening the screw tends to close the loop rather than to open it. Tighten the screw till it makes contact with the wire, then tighten it about another half-turn, to squeeze the wire a bit. Never make the errors shown in Fig. 4-14. The insulation should end not more than one quarter inch from the screw, at the most. Note that these two illustrations, reproduced by courtesy of Underwriters Laboratories, were made specifically for aluminum wire, but the principles are correct for copper wire also.

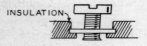

Fig. 4-11. Terminal on a typical receptacle, or switch.

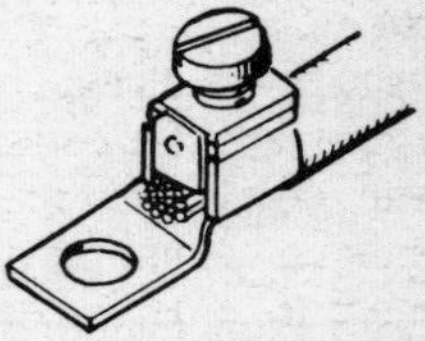

Fig. 4-12. Solderless connectors of this type are used with heavy sizes of wire.

There are times when it might appear logical to connect two wires under a single wrap-around terminal screw. Don't do it; the Code prohibits it in Sec. 110-14(a). Take those two wires and another short length of the same wire, and connect all three together using a wire nut described later in this chapter. See Fig. 4-15. Then connect the remaining end of the short wire under the terminal screw. Most connectors of the kind shown in Fig. 4-12 are for one wire, unless marked to indicate the number and size of wires it will accommodate. The marking could be on the carton if the connector itself is too small for the marking.

How long should the bare wire be, before you connect it under the screw? Professionals leave about three inches or so of bare wire, and after the terminal screw has been tightened, there is a "tail" a few inches long beyond the screw. Twist this tail a few times and it will break off near the screw. Another way is to leave just enough bare wire to go around the screw, form a loop with a pair of long-nose pliers, slip it around the screw and then close it with the same pliers so that the loop is entirely under the screw, before tightening the screw.

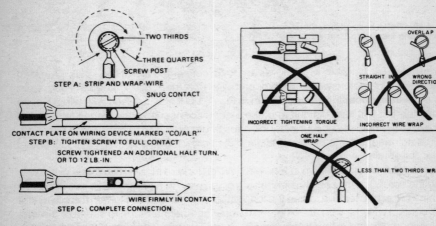

Fig. 4-13. Be sure wire is wrapped around terminal screw in clockwise fashion, so that tightening the screw tends to *close* the loop.

Fig. 4-14. Avoid these common errors when connecting wires to terminal screws.

Many 15-amp. and 20-amp. receptacles and switches have no terminal screws at all. Instead, there are internal clamps that grasp a straight piece of wire pushed into them, forming an effective connection. Merely strip the end of the wire for half an inch or so (the proper length is usually shown on the device itself), and push the wire into the hole on the device. See Fig. 4-16. If an error is made, remove the wire by pushing a small screwdriver blade into another opening on the device. But do note that while these push-in connections are acceptable for copper or copper-clad aluminum wire, they may *not* be used with all-aluminum wire.

In conduit wiring, sometimes a wire is pulled through one box and on to another, possibly through still another, and so on. If the wire merely runs through the box, pull it through without splice, but leave a loop in case it is necessary to repull it in the future. If you intend to make a connection to the wire as it passes through the box, let a loop several inches long project out of the box. Strip away an inch or so of insulation, form a loop, and connect it under one terminal screw as shown in Fig. 4-17.

Wires larger than No. 10 may not be connected under a terminal screw; use a connector of the general type shown in Fig. 4-12.

Splices: The splices must be electrically as good as an unbroken length of wire. The insulation of the splice must be as good as that on the original wire, which is accomplished when using insulated solderless connectors properly installed. When two or more ends of wire must be connected to each other, lay the wires together as was shown in Fig. 4-15, and install a solderless connector or "wire nut" of the general type shown in Fig. 4-18. One type has a threaded metal insert molded into the insulating shell; screw the connector onto the wires to be joined. The other kind has a removable metal insert. Slip the insert over the wires to be joined, tighten the screw of the insert, then screw the insulating shell over the metal insert.

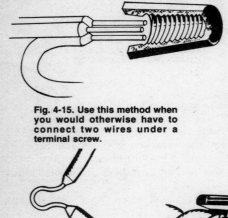

Fig. 4-15. Use this method when you would otherwise have to connect two wires under a terminal screw.

Fig. 4-16. Strip the wire, push it into opening on switch or receptacle, and the connection is made.

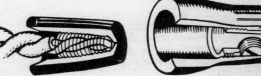

Fig. 4-17. In conduit wiring, a continuous wire may be connected to a terminal as shown here.

Fig. 4-18. Ordinary splicing is simple when using solderless connectors ("wire nuts").

The spring-loaded connector shown in Fig. 4-19 is also popular. Inside its insulating shell there is a cone-shaped metal spring. Screw the connector over the wires to be joined; the insulating cover provides a good grip. In screwing it on, the coil spring temporarily unwraps, but when released forms a very tight grip on the wire.

Fig. 4-19. This kind of connector contains a tapered coil spring inside the insulating cover.

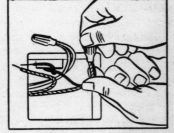

In using these connectors, if one wire is much smaller than the others, let it project a bit beyond the heavier wires. If you have removed the right length of insulation from each wire, the insulating shell will cover all bare wires and no taping is necessary. Note that all styles of these connectors are available in various sizes, depending on the number and size of the wires to be joined.

For wire sizes No. 10 and smaller, the insulated "clamshell" connector shown in Fig. 4-20 may be used. The wire insulation is used to position the wire in this type connector, so do *not* strip the wire insulation before installing in the connector. A squeeze with parallel-jaw pliers installs the self-insulating connector.

Fig. 4-21. Crimp type wire connector, and pliers with special crimping die in handle. For copper wire only.

Fig. 4-20. "Clamshell" wire connector for taps or pigtails, copper wire only.

Another wire connector, for No. 10 and smaller wire, is the shell shown in Fig. 4-21. Unless the manufacturer's instructions direct otherwise, first twist the wires together, then slip the shell over them and crimp, using a tool of the type shown. Then cut off the wire ends which could puncture the insulation, and either tape or use formed plastic caps to insulate.

Important: Whichever type of wire connector you use, be certain it is listed for the number and size of wires you have to join.

For wires that are too large to be joined by the connectors described, use heavy duty copper connectors of the style shown in Fig. 4-22.

Sometimes a wire must be spliced to another *continuous* wire. In the heavier sizes, the simplest way is to use one of the split-bolt connectors shown in Fig. 4-23. Tape after making the connection; some connectors are available with an insulating cover that can be snapped on after making the connection. For smaller sizes, it is usually simpler to cut the continuous wire, thus forming two ends; the wire to be spliced in the connection makes the third wire. Then use a solderless connector or wire nut, as was shown in Fig. 4-15.

Fig. 4-23. Use this split-bolt connector when splicing a heavy wire to another continuous heavy wire.

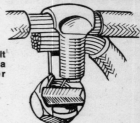

Fig. 4-22. For heavier wires, use metal connectors. They must be taped.

Insulation: Splicing devices such as the wire-nut are inherently insulating. Others have available insulating covers or boots which can be added after the splice is made. Some, like those shown in Figs. 4-22 and 4-23 must be taped. Use electrician's plastic tape which has a very high insulating value in spite of its being quite thin. Starting well back on the wire insulation, wrap the tape on spirally, partially overlapping each turn, from one end to the other, and then back with the spiral in the opposite direction. If the splice will be subject to mechanical strain or abuse, apply additional layers, as a "cushion."

Chapter 5

CIRCUIT BREAKERS, FUSES AND CIRCUITS

The Code has established the ampacity, or maximum amperage which is considered safe for any particular kind and size of wire. This maximum for *copper* wires, is shown in the Table in Chap. 4. (For other sizes, and for *aluminum* wire, see Tables 310-16 through 310-19 in your copy of the Code.) If more than the permissible maximum amperage is allowed to flow, the temperature of the wire goes up, and the insulation may be damaged, leading to shortened life and grounds that can become dangerous. If the overload is great enough, there is danger of fire.

Overcurrent Devices: The amperage in any wire is limited to the maximum permitted by using what the Code calls an overcurrent device. Two types are in common use: fuses and circuit breakers, both rated in amperes. Any overcurrent device you use must have a rating in amperes, not greater than the ampacity of the wire which it protects. For example, No. 12 wire has an ampacity of 20 amp., therefore the circuit breaker or fuse that you use to protect the wire must have a rating not greater than 20 amp.

When two different sizes of wire are joined together (as, for example, when No. 8 is used for mechanical strength in an overhead run, joined to No. 14 for the inside wiring of the building to which the wire runs) the overcurrent protection must be of the right size for the *smaller* of the two wires. Of course a size correct for the *larger* wire may be used at the starting point, provided another one of the proper size for the *smaller* wire is used where the wire is reduced in size. This is often done when a run of say No. 8 wire with ampacity of 40, runs from one building to another, where in turn it feeds several 15-amp. circuits, all as shown in Fig. 5-1.

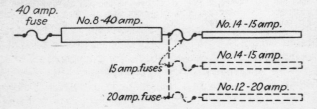

Fig. 5-1. If fuses are used where the wire size is reduced, select a size that protects the smaller wire.

Fuses: A fuse is nothing more or less than a short piece of metal, of a kind and size which experiment has shown will melt when more than a pre-determined number of amperes flows through it. This metal link is enclosed in a convenient housing, to prevent hot metal from spattering if the fuse blows, and to permit easy replacement. A fuse rated at 15 amp. is tested to carry 15 amp. continuously. When more than 15 amp. flow through it, the wire inside the fuse melts (the fuse "blows"), which is the same as opening a switch or cutting the wire. The greater the overload, the quicker the fuse blows. The most ordinary fuse is the plug type, shown in Fig. 5-2, made only in ratings of 30 amp. and less. It is known as the Edison-base type; its base is the same as on ordinary lamps.

Time-Delay Fuses: An ordinary fuse carries 80% of its rated amperage indefinitely, but blows very quickly if twice that amperage flows through it. For that reason fuses often blow when a motor is started, because a motor which draws only 6 amp. *while running* may draw as much as 30 amp. for a few seconds *while starting*. On the other hand, wire which can safely carry 15 amp. continuously (and which might be damaged or even cause a fire if 30 amp. flowed continuously) will not be damaged in the least, or cause a fire, if 30 amp. flow for a few seconds. For this reason another type fuse was developed, known as the "time delay" type. It is commonly called by its trade name of "Fusetron" although there are other brands. It blows just as quickly as an ordinary fuse on a small *continuous* overload, or on a short circuit, but *it will carry a big overload safely* for a fraction of a minute. This type of fuse is rapidly becoming popular, especially where motors are used. It prevents needless blowing of fuses and does away with many service calls. It looks like an ordinary fuse, but is made differently inside.

Important: Edison-base fuses, whether the ordinary type shown in Fig. 5-2 or the time-delay "Fusetron," are no longer permitted in new installations. They may be used only as replacements, and then only if there is no evidence of overfusing or tampering.

Fig. 5-2. Plug fuses are not made in ratings over 30 amp.

Fig. 5-3. A typical Type-S non-tamperable fuse, and its adapter. Once an adapter has been screwed into a fuse-holder, it cannot be removed. This prevents use of fuses larger than originally intended.

Non-Tamperable Fuses: Since all ratings of ordinary Edison-base plug fuses are interchangeable, nothing prevents one from using, for example, a 25- or 30-amp. fuse to protect a No. 14 wire, which has an ampacity of only 15 amp. However, this is unsafe. For that reason there was developed a non-tamperable fuse, which is shown in Fig. 5-3. The Code calls this a "Type S" fuse; it is commonly called by its trade-name of "Fustat." The fuse itself will not fit an ordinary fuse-holder, so an adapter shown in the same illustration must first be installed in the ordinary fuse-holder; once installed, it cannot be removed. There are three sizes of adapters. The 15-amp. will accept only 15-amp. or smaller fuses; the 20-amp. will accept only 16 through 20-amp. fuses; the 30-amp. will accept 21 through 30-amp. fuses. This is a safety measure. The Type S fuse is presently made only in the time-delay type. The Code requires the use of Type S fuses in all *new* installations, and for replacements if there is evidence of tampering or overfusing.

One word of caution is in order when using Fustats: when screwing the fuse into its holder, turn it some more after it appears to be tight. Under the shoulder of the fuse there is a spring that must be flattened. Unless you screw the fuse in very tightly, enough to flatten this spring, the fuse will not "bottom" and you will have an open circuit: the symptoms will be the same as if the fuse were blown. Plug fuses are rated at 125 volts, but may be used on a system having a grounded neutral and no conductor over 150 volts to ground, so could be used for a 240 volt load served from a 120/240 volt 3-wire system.

Fig. 5-4. Cartridge fuses rated 60 amps. or less are of the ferrule type shown.

Fig. 5-5. Cartridge fuses rated more than 60 amp. have knife-blade terminals shown.

Cartridge Fuses: This type is made in all amperage ratings. Those rated at 60 amp. or less are of the ferrule type shown in Fig. 5-4; those rated at 70 amp. or more have knife-blade terminals as shown in Fig. 5-5. The most ordinary type may be used in any circuit of not over 250 volts between conductors.

Circuit Breakers: More and more circuit breakers are being used in place of fuses. A circuit breaker looks something like a toggle switch, with a handle that lets it be used just like a switch to turn power on and off. See Fig. 5-6 which shows a single unit. Inside each breaker is a fairly simple mechanism which in case of overload trips the breaker and disconnects the load. If a breaker trips because of overload, in most brands you must force the handle *beyond* the OFF position, then return it to ON, to reset it. On some brands however, the handle merely goes to the OFF position; reset it by returning it to the ON position.

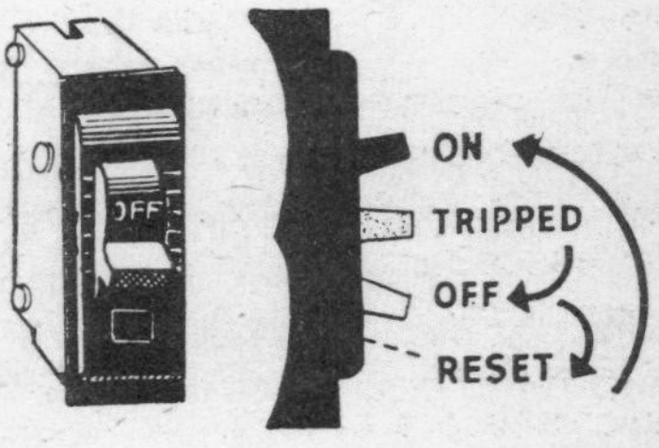

Fig. 5-6. A single circuit breaker, and method of resetting the breaker if it trips.

A circuit breaker has a definite time delay. It will carry 80% of its rated load indefinitely, a small overload for a considerable time, and trip quickly on a large overload. Nevertheless it will carry temporary overloads long enough to permit motors to start.

When a Fuse Blows: What is to be done when a breaker trips or a fuse blows? Most people will say: reset the breaker, or install a new fuse. Correct, but first find out why the fuse blew. Fuses are the safety valves of electrical installations. To use substitutes, or fuses that are too large for the size of the wire involved, is either ignorance or stupidity. Don't do it, any more than you would drive an automobile without brakes.

When a circuit breaker trips or a fuse blows, it is for one of two reasons. Either something connected to the circuit is defective thus drawing an excessive number of amperes, or there are too many lights, appliances or motors connected at the same time, thus overloading the circuit.

If a fuse blows quickly every time a particular appliance is plugged in, especially if it makes no difference whether it is plugged in downstairs or upstairs on a different circuit, the appliance is defective. Often the defect is in the cord. If the appliance has a removable cord, try a different cord; if the cord is permanently attached, only careful inspection will locate the defect. Badly twisted and worn cords must be replaced, *not repaired.*

If the fuse blows when a motor (as in a home workshop) is turned on, remember that a motor which consumes only 6 amp. while running, may draw over 30 amp. for a few seconds while starting. That may be the reason for the fuse blowing. Substituting a time-delay fuse for the ordinary fuse may solve the problem. If however the fuse continues to blow, suspect the motor. Check the cord. Check to see that the motor bearings have oil. (Some motors have sealed bearings never requiring oil.) Perhaps the belt is too tight, or the machine that the motor drives lacks oil, increasing the load on the motor, and increasing the amperes beyond the safe point.

If fuses blow or circuit breakers trip from time to time on any one circuit, for no apparent reason, it is likely that the circuit is just simply overloaded. There is no choice except to disconnect some of the load on that circuit. The wise procedure would be to install an additional circuit.

Branch Circuits: If all the lights and appliances in a home or on an entire farm were protected by a single fuse or circuit breaker, the entire establishment would be in darkness when that breaker trips or the fuse blows. That would be most inconvenient. Also, all the wires would have to be very large to match the ampere rating of that breaker or fuse, and that would mean a clumsy and expensive installation. Therefore the different outlets* in an installation are separated into smaller groups known as branch circuits. Inexpensive No. 14 or No. 12 wire is used for most of the wiring, protected by 15- or 20-amp. breakers or fuses. When one of these breakers trips, or a fuse blows, only those outlets on that circuit are dead; those on other circuits are still alive.

Continuous Loads: See discussion under Water Heaters in Chapter 14.

Outlets per Circuit: In residential work, you may place as many lighting outlets as you please on one circuit, so far as the Code is concerned. On the other hand if you put too many on one circuit, you will probably have trouble with breakers tripping or fuses blowing. In most cases you will not want to connect as many as a dozen outlets on one circuit, even if they are permitted.

---

* Every point where electric power is actually used is an outlet. Each fixture, even if it has five lamps, is considered one outlet. Each receptacle outlet, even if it is the duplex type, is one outlet. Switches are not outlets because they use no electric power, but merely control its use. However, in estimating the cost of an installation, switches are included in counting the outlets, to arrive at a total cost on the "per outlet" basis, but this usage is not in accordance with the Code definition.

Kinds of Circuits: The circuits used in homes can be divided into three general types, each of which will be discussed in detail later in this chapter. These are:

1. Lighting Circuits: These are primarily for lighting and serve permanently installed lighting fixtures, as well as receptacle outlets into which you plug lamps, radios, televisions, clocks, and similar 120-volt devices (but not kitchen appliances).

2. Small Appliance Circuits: Receptacles in kitchen, dining room, etc.

3. Individual Appliance Circuits: These are circuits each serving a single appliance such as a range, water heater, automatic clothes washer or dryer. The circuit may be either 120-volt or 240-volt.

Lighting Circuits: The Code requires enough *lighting* circuits so that 3 watts of power will be available for every square foot of floor space in the house. Since a circuit wired with No. 14 wire and protected by 15-amp. overcurrent protection provides 15 × 120, or 1800 watts, each circuit obviously is enough for 1800 ÷ 3, or 600 square feet.

Do note however that this is the Code *minimum.* The Code concerns itself not with convenience or utility, but with *safety* only. Following the Code suggestion of one circuit for every 600 sq. ft. is likely to result in an installation that may not be entirely acceptable to the owner over a period of time. You will be wise if you provide one circuit for each 500 sq. ft. of space.

How are the square feet of space in a house figured? Open porches and garages even if attached to the house are not included in the total. Unfinished spaces and unused spaces *if adaptable for future use,* positively *must* be included. This is a wise Code provision, for in the past in too many cases, such spaces intended to be finished later were provided with perhaps a single outlet on an already loaded circuit. When the space was later finished off, cable from this one outlet was led off to serve half a dozen new outlets, hopelessly overloading the first circuit.

Today, such space must be counted in the total area. Run a special circuit to the area, terminate it in a single outlet if you wish, and later branch off from that outlet to the additional outlets that you want when you finish off the space. If the space is large, run two circuits.

To arrive at the Code definition of the total number of square feet for determining the number of circuits, simply multiply the length of the house by its width. If its outside dimensions are 30 × 45 feet, one floor represents 30 × 45 or 1350 sq. ft. If the basement is finished or can be finished into usable space, add its area of, for instance, 24 × 30, or 720 sq. ft. to the first-floor area of 1350 sq. ft., for a total of 2070 sq. ft. If an upper floor has unfinished space that can later be finished into a bedroom, add its area.

Then divide the total area by 600 to arrive at the minimum number of lighting circuits the Code requires, or divide by 500 for a more adequate installation. All this can be reduced to a table, as shown on the following page.

Special Small Appliance Circuits: The circuits already discussed are for lighting, including floor and table lamps, and minor appliances such as radio, TV, vacuum cleaner, and the like. These circuits will not permit proper operation of larger kitchen and similar appliances, which consume much more power. The Code in Sec. 220-3(b) (1) requires two special *small appliance* circuits to serve only small appliance outlets including refrigeration equipment in kitchen, pantry, breakfast room, dining room and

No of sq. ft.	Number of Lighting Circuits	
	Code Minimum	Recommended
1000	2	2
1200	2	3
1600	3	4
2000	4	4
2400	4	5
2800	5	6
3200	6	7

family room. Both circuits must extend to the kitchen; the other rooms may be served by either one or both of them. The circuits must be wired with No. 12 wire, protected by 20-amp. circuit breakers or fuses. No lighting outlets may be connected to these circuits*. But do note that either 15- or 20-amp. receptacles may be installed on these 20-amp. circuits.

Each such circuit has a capacity of 20 × 120 or 2400 watts, which is not too much considering that toasters, irons and similar appliances often require 1000 watts; some roasters and other appliances consume over 1500 watts. These two circuits can well be merged into one 3-wire circuit, which will be discussed in Chap. 13.

At least one separate 20-amp. circuit must be run to the laundry, for washers, irons and so on. A clothes dryer requires its own circuit, usually 30-amp.

Individual Appliance Circuit: It is customary to provide a separate circuit for each of the following appliances:

1. Self-contained range
2. Separate oven, or
   counter-mounted
   cooking unit
3. Water heater
4. Automatic clothes washer
5. Clothes dryer
6. Garbage disposer
7. Dishwasher
8. Motor on oil-burning furnace
9. Motor on blower in furnace
10. Water pump
11. Permanently connected appliances
    rated at more than 1000 watts
    (for example, a bathroom heater)
12. Permanently connected motors
    rated more than 1/8 hp.

Note that such circuits may be either 120-volt or 240-volt, depending on the particular appliance or motor installed. These circuits will be discussed in more detail in Chap. 14.

Total Number of Circuits: You will have to decide for yourself how many circuits you will need. Provide circuit breakers or fuses for specific appliances, even if you do not intend to install such appliances until later. Even then, provide for some spare circuits.

---

* An electric clock receptacle (shown in Fig. 13-2) in a kitchen or dining area that not only serves but also supports a clock, may be connected to either a lighting circuit or one of the small appliance circuits.

For a really modern installation, follow the suggestions of the table below, which is based on one lighting circuit for each 500 sq. ft. of area.

No. of square ft. of area in house	1000	1200	1600	2000	2400	2800	3200
120-V. CIRCUITS:							
General-purpose circuits (one per 500 sq. ft.)..	2	3	4	4	5	6	7
Ditto, spares	1	1	1	2	2	2	2
Small appliance cct.	2	2	2	2	2	2	2
Special laundry circuit.....................	1	1	1	1	1	1	1
Individual appliance ccts., 120-v.							
Oil or gas burner........................	1	1	1	1	1	1	1
Blower on furnace.......................	1	1	1	1	1	1	1
Bathroom heater, room air conditioner,							
workshop motor, etc....................	—	—	—	—	—	—	—
Spare....................................	1	1	1	2	2	2	2
TOTAL 120-V. CCTS.	—	—	—	—	—	—	—
240-V. CIRCUITS:							
Individual appliance circuits:							
Range	1	1	1	1	1	1	1
Water heater	1	1	1	1	1	1	1
Clothes dryer	1	1	1	1	1	1	1
Water pump, etc.........................	—	—	—	—	—	—	—
Spare....................................	1	1	1	1	1	1	1
TOTAL 240-V. CCTS.	—	—	—	—	—	—	—

Chapter 6

CIRCUIT BASICS

Before you can actually wire a building, you must first learn how switches, receptacles, sockets and other devices are connected to each other with wire, to make a complete electrical system.

Grounded Wires: In residential and farm wiring, one of the current-carrying wires is grounded, that is, connected to an underground metal water pipe system, and to a driven ground rod. This grounded wire is called just that: the grounded wire. (Many people call the *grounded* wire the *neutral* wire. Sometimes it is, sometimes it is not. All this will be explained in another chapter.) All ungrounded wires are known as "hot" wires. It is most important that you remember the following four points:

1. The grounded wire is *always* white in color.
2. The grounded wire must run direct to every 120-volt device to be operated (*never* to anything operating only at 240 volts).
3. The grounded wire is *never* fused, or protected by a circuit breaker.
4. The grounded wire is *never* switched or interrupted in any other way.

With one exception (discussed in Chap. 10) white wire may never be used except as a grounded wire. Other wires are usually black but may be some other color, but not white (with the one exception just mentioned) or green.

The white wire must run to every 120-volt device *other than a switch;* the other wire is usually black but may be some other color, but not white or green. Examine a socket or similar device carefully and you will find that one of its two terminals is a natural brass color, the other is a white color, usually nickel-plated or tinned. *The white wire must always run to the white terminal.* In the case of sockets, the white terminal in turn is always connected to the screw shell and never to the center contact of the socket. Switches never have white terminals.

Wiring Diagrams: Two wires must run from the starting point to each outlet whether it serves a lamp, motor, or any other electrical equipment. Imagine, if you wish, that the current flows out over the black wire to the equipment, through the equipment, and back to the starting point over the white wire. We will call the starting point the SOURCE, and use a lamp to indicate the device to be operated. We will use the symbol of Fig. 6-1 instead of a picture of a lamp. Note also the method shown in Fig. 6-2 to indicate

Fig. 6-1. This symbol indicates a lamp in all diagrams.

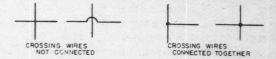
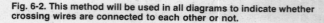

CROSSING WIRES NOT CONNECTED   CROSSING WIRES CONNECTED TOGETHER

Fig. 6-2. This method will be used in all diagrams to indicate whether crossing wires are connected to each other or not.

whether crossing wires are connected to each other or not. In all diagrams the white wire will be indicated by a light line like this ——; black wire by a heavy line like this ▬; the black wire between a switch and the outlet which it controls by a heavy broken line like this - - -. The fact that one line is heavier than the other does not mean that one wire is larger than the other; both are the same size.

The simplest possible diagram is that of Fig. 6-3: a lamp which is always on, with no way of turning it off. Such a circuit is of little value, so, add a switch. This is very simple to do; all you do is to make the *black* wire detour to a switch as shown in Fig. 6-4. An unenclosed switch is shown so that you can see how it operates; opening the switch is the same as cutting a wire. Of course unenclosed switches are unsafe and not permitted. Fig. 6-5 shows the same circuit but with an enclosed flush toggle switch which serves the same purpose but is far neater and more convenient and, having no exposed mechanism, is safer.

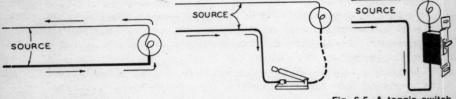

Fig. 6-3. The simplest circuit: a lamp always on.

Fig. 6-4. The circuit of Fig. 6-3, plus an open switch.

Fig. 6-5. A toggle switch (closed) substituted for the open switch in Fig. 6-4.

If several lamps are to be used, do not make the mistake of wiring them "in series," as the wiring shown in Fig. 6-6 is known, which is impractical except for very special purposes. For example, if one lamp burns out, all go out (as in the very old style "series" type of Christmas tree lights, which a few "old-timers" will remember).

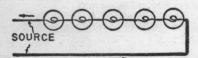

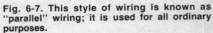

Fig. 6-6. This style of wiring is known as "series" wiring; it is impractical except for very special purposes.

Fig. 6-7. This style of wiring is known as "parallel" wiring; it is used for all ordinary purposes.

Wire them "in parallel," as the scheme shown in Fig. 6-7 is known. The white wire goes to each lamp, from the first lamp to the second and to the third and so on; the black wire also runs to each lamp in similar fashion. Each lamp will still light even if one or more of the others is removed. The switch shown cuts off the current to all the lamps at the same time. The wiring is most simple: white wire from SOURCE to each socket or outlet; black wire from SOURCE to the switch, and from the switch another black wire to each of the sockets. This is the way a fixture with 5 sockets is wired.

If each lamp is to be controlled by a separate switch, use the diagram of Fig. 6-8. The white wire always runs from SOURCE to each lamp, the black wire from SOURCE to each switch; from each switch a black wire runs to the lamp which it is to control. Trace the current in over the black wire through each switch separately, along the black wire to the lamp, and back from lamp to the SOURCE. Trace it carefully and you will find that each lamp can be separately controlled.

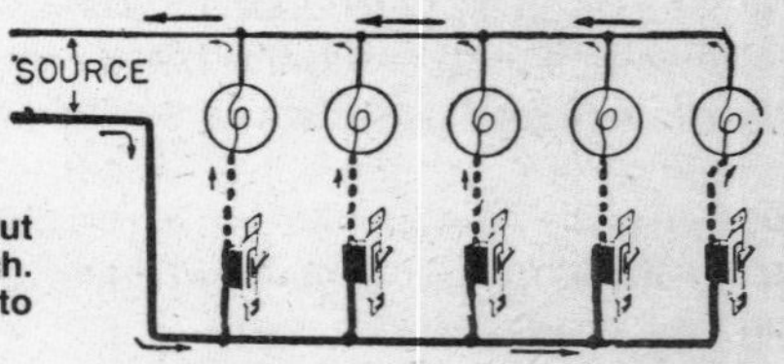

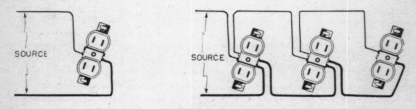

Fig. 6-8. The same diagram shown in Fig. 6-7, but each lamp is now controlled by a separate switch. Note that the white wire runs directly from SOURCE to each lamp.

Wiring Receptacle Outlets: The wiring of plug-in receptacle outlets is most simple. If there is only one receptacle, run the white wire from SOURCE to one side of receptacle that has whitish terminal screws, the black to the other, as in Fig. 6-9. If there are several receptacles, run the white wire to one side of the first receptacle, from there to the second, and so on; do the same with the black. See Fig. 6-10; study this picture carefully and you will see that it is the same as the diagram of Fig. 6-7, except that receptacles have been substituted for lamps, and the switch omitted.

Fig. 6-9. The wiring of receptacles is most simple.

Fig. 6-10. Receptacles usually have two terminals on each side, to make it easy to run wires from one to the next.

However, as was mentioned in Chap. 2 and will be discussed in more detail in Chap. 9, many times you will want to have a receptacle controlled by a switch. The Code requires that every habitable room, and every hallway, stairway, outside entrance, and attached garage, must have some switch-controlled lighting. This lighting must come from permanently-installed fixtures in bathrooms and kitchens, but could be from portable lamps plugged into switch-controlled receptacles in other rooms.

Your receptacles will be the grounding type which will be described in Chap. 7, and which require a third wire to be run to the *green* terminal of the receptacle. However in *this* chapter we will disregard that third wire, consider only the two current-carrying wires as on old style receptacles. How to connect the third wire to the green terminal will be discussed in later chapters, for the details vary with the particular wiring method you use, whereas the two current-carrying wires are connected the same, regardless of wiring method.

Wiring Outlets Controlled by Pull-Chain: Since the pull-chain mechanism is part of the socket, an outlet that is to be controlled only by pull-chain is wired exactly like a receptacle outlet. Simply substitute the pull-chain socket for the receptacle in either Fig. 6-9 or 6-10. Attics and closets are typical locations where a pull-chain fixture might be used.

Combining Several Diagrams: When several groups of outlets are to be wired, the two separate wires from each group can be run back to a common starting point, where the wires enter the house, as is shown in Fig. 6-11 where the diagrams of Fig. 6-5, 6-7, and 6-8 have been combined into one. This would require a very considerable amount of material, and the wires would be much longer than necessary. It is much simpler to merely wire the first group, then to run wires from the first group to the second, and from the second to the third, as shown in Fig. 6-12, which shows the same three groups but using less material.

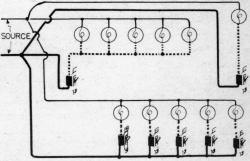

Fig. 6-11. The outlets of Figs. 6-5, 6-7 and 6-8 have been combined into a single group of outlets, in other words here form one circuit. It still uses a great deal of material; to reduce the amount required, the circuit can be arranged as the next diagram shows.

The *white* wire may be extended at any point whatever to the next outlet. The black wire likewise may be extended at any point, provided only that it can be traced all the way back to the SOURCE without interruption by a switch. In other words, in these diagrams a black wire can be extended from any black wire indicated by a solid heavy line like this ————, but not from the black wire between a switch and an outlet indicated by a broken line like this – – –. Thus, in Fig. 6-12, *A* and *B* are the starting points for the second group, and *C* and *D* are the starting points for the third group of lights.

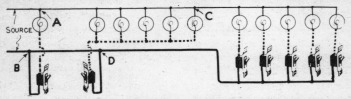

Fig. 6-12. The diagram of Fig. 6-11 rearranged to use less material.

3-Way Switches: The switch used for controlling a light from one point is known as a "single-pole" switch. Often you will want to turn a light on or off from two different places, as for example, a hall light from upstairs and downstairs, or a garage light from house and garage, or a yard-light from house and barn. For this purpose, switches known as the "3-way" type are used. Despite their name, they will not control a light from three different places, but only two. Such switches have three different terminals for wires, and their internal construction is similar to that shown in Fig. 6-13. In one position of the handle, terminal *A* is connected inside the switch to terminal *B*; in the other position of the handle, terminal *A* is connected to terminal *C*. *The common terminal A is usually identified by being of a darker color than the others, which are natural brass.*

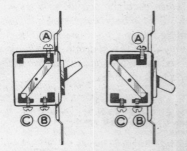

Fig. 6-13. This shows what happens inside a 3-way switch when the handle is thrown from one position to another. The current enters by the "common" terminal A, leaves by either B or C depending on the position of the handle.

Study the diagram of Fig. 6-14, in which the handles of both switches are down. The current can be traced from SOURCE over the black wire through switch No. 1 through terminal *A*, out through terminal *B* and up to terminal *C* of switch No. 2, but there it stops. The light is out.

Next see Fig. 6-15 where the handles of both switches are up. The current can be traced from SOURCE over the black wire through switch No. 1 through terminal *A*, as before, but this time out through terminal *C*, and from there up to terminal *B* of switch No. 2. There it stops and the light is out.

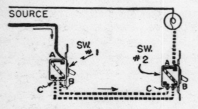

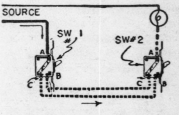

Fig. 6-14. A light controlled by two 3-way switches. With both handles down, the light is out.

Fig. 6-15. With the handles of both switches up, the light is still out.

Now examine the diagrams of Fig. 6-16 and 6-17, in both of which the handle of one switch is up, the other down. In each case, the current can be traced from SOURCE over the black wire, through both switches, through the lamp, and back to SOURCE. The light in either case is on. In the case of either Fig. 6-16 or 6-17, throwing either switch to the opposite position changes the diagram back to either Fig. 6-14 or 6-15, and the light is out. In other words, the light can be turned on and off from either switch.

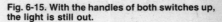

The wiring is simple, as these diagrams show. Run the white wire from SOURCE as usual to the light to be controlled. Run the black wire from SOURCE to the common or marked terminal of the *first* 3-way switch. Run a black wire from the common or marked terminal of the *second* 3-way switch to the light. That leaves two unused terminals on each switch; run two black wires from the terminals of the first switch, to the two terminals of the second switch. It makes no difference whether you run a wire from *B* of the first switch to *C* of the second as shown, or from *B* of the first to *B* of the second. The wires that start at one switch and end at another are called runners, travelers, or jockey legs.

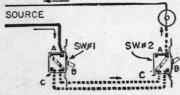

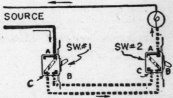

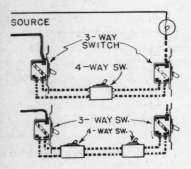

Fig. 6-16. With one handle up, the other down, the light is now on. Trace the current along the arrows.

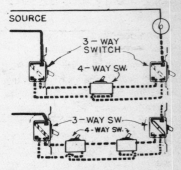

Fig. 6-17. Both handles are now the opposite of Fig. 6-16, but the light is still on. Either switch controls it.

4-way Switches: If a light is to be controlled from more than two points, use two 3-way switches, and 4-way switches at each remaining point, whether there are three, four or more points in total. A 4-way switch can be identified by the fact that it has four terminal screws and *no* "on"-"off" markings on the handle. Install the 3-way switches at the points nearest the SOURCE and nearest the light, with 4-way switches in between, all as shown in Fig. 6-18.

Fig. 6-18. Usually four-way switches are connected as shown above.

Fig. 6-19. Some brands of 4-way switches are connected as shown above.

Different brands of 4-way switches are made in two different ways, so that sometimes the diagram of Fig. 6-18 is not correct, and you must use that of Fig. 6-19 instead. If you are not sure of the type you have, try one diagram; if it does not work, try the other. You can do no harm by wrong connections, except that the circuit will not work if you have the wrong connection.

Pilot Lights: When a switch controls a light that can't be seen from the switch, the light is often left turned on when it should be turned off. Install a pilot light (a small, low-wattage lamp) near the switch, so that both lights are turned on and off at the same time. The pilot light then serves as a reminder that the other light (which can't be seen from the switch) is on. Pilot lights are commonly used in connection with basement and attic lights in homes, with hay-mow lights in barns, and similar locations. Their wiring is very simple as Fig. 6-20 shows. First consider the diagram, disregarding the wires shown on dotted lines, as well as the pilot light itself. It is then the same as Fig. 6-5. Add the wires shown in dotted lines and it becomes the same as the diagram of Fig. 6-7 except that there are only two lamps instead of five. The white wire runs to both lamps; the black wire from the SOURCE runs to the switch as usual; the black wire from switch runs to both lamps, and the diagram is finished.

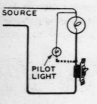

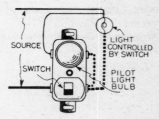

Fig. 6-20. The basic circuit of a pilot light.

Fig. 6-21. The wiring scheme of a combination pilot light and switch. Another brand might be wired somewhat differently.

For this purpose, a combination switch and pilot light similar to that shown in Fig. 6-21 is often used; the pilot lamp and the switch are combined in one device. The diagram shown in connection with this picture is correct only for some brands; exact diagrams usually come with each device. There are also switches having a small lamp in the handle which serves as a pilot light (or, in a slightly different variety, a light that is on while the switch is in the "off" position, to aid in finding the switch in the dark).

Other Diagrams and Combinations: You will have little difficulty making a diagram for any desired combination of outlets and switches, whether single-pole or 3-way, if you will bear in mind the principles covered in this chapter. To make any diagram, first locate each outlet where desired and run the white wire (light line like this ——————) from SOURCE to every outlet. Then run a black wire (heavy line like this ——— from SOURCE to each outlet that is not to be controlled by a switch. Next run black wire (heavy broken line like this - - -) from each outlet which is to be controlled by a switch, to the switch which is to control it; if 3-way switches are involved, run additional ones between the switches which are to control the outlet. Finally, from each switch run black wire (heavy line like this ——————— back to SOURCE. You may have to rearrange some of the wires when you have finished, to reduce the amount of wire used, but this will not affect the proper operation of the hook-up.

Chapter 7

IMPORTANCE OF GROUNDING

In all discussions concerning electrical wiring, you will regularly meet the terms ground, grounded, grounding. They all refer to deliberately connecting parts of a wiring installation to a grounding electrode or electrode system. Grounding falls into two categories: (a) System grounding, or grounding one of the current-carrying wires of the installation, and (b) Equipment grounding, or grounding non-current-carrying parts of the installation, such as the service equipment* cabinet, the frames of motors or electric ranges, the metal conduit or armor of armored cable, and so on.

The purpose of grounding is *safety.* If an installation is not properly grounded, it can be exceedingly dangerous as to shocks, fires, and damage to appliances and motors. Proper grounding reduces such dangers, and also minimizes damage from lightning, especially on farms. Grounding is a most important subject; it is so important that many points will be repeated throughout this book, for emphasis. Study the subject thoroughly; *understand it.*

The Code rules for grounding are quite complicated, and at times appear to be ambiguous. However, for installations in homes and farm buildings (the only areas covered in this book) they are relatively simple. In this chapter will be discussed only the basic principles of grounding; other chapters will discuss sizes of the ground wire, ground clamps, ground rods, and similar details.

Grounded and "Hot" Wires: Three wires run from the power supplier's transformer to the building to be served. One of the wires is grounded at the transformer, and also at the service equipment of the building. All this is shown in Fig. 7-1 where the wires are labeled *A, B,* and *N.* The wire *N* is grounded and is called just that: the grounded wire. Wires *A* and *B* are called hot wires. The voltage between *A* and *N,* or *B* and *N,* is 120 volts; between *A* and *B* it is 240 volts.

Fig. 7-1. Using only three wires, two different voltages are available. Use the lower voltage or 120 volts for low-wattage devices, the high voltage or 240 volts for high-wattage items such as ranges and water heaters. Note the symbol for a connection to ground.

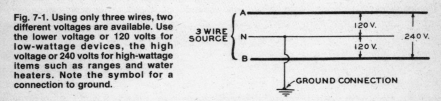

Between the transformer and the service equipment, the grounded wire *N* is a *neutral* wire. In wiring a building, any wire connected to the point where the neutral wire ends in

* In any installation there is either a main fused service switch, or a circuit breaker panelboard cabinet serving the same purpose. When in this book the reference is to "service equipment" it will refer to either of the two, whichever is installed.

the service equipment is (with an exception in connection with *3-wire* circuits that will be discussed in Chap. 13) just a *grounded* wire. Many people call it a neutral but it is not a neutral; there *cannot* be a neutral in a *2-wire* circuit.

The grounded wire must be white, with a few exceptions that will be discussed later, where it can be bare. The hot wires may not be white or green; if there is only one, it is black; if there are two they are usually black and red, but two blacks could also be used.

Terminology: In this book, "ground" means a metal water pipe bonded to a driven ground rod, or where the water pipe is not available, to the rod alone. The Code calls this a grounding electrode system, or the rod alone a grounding electrode. The verb "to ground" something, means to connect it to the ground.

The term "bonded" or "bonding" means connecting together two or more points of non-current-carrying parts (conduit, boxes, etc.), often with bare, uninsulated wire, so that *all* such parts become connected together, and finally to the ground.

The term "ground*ed* wire" means that wire of a circuit, normally carrying current, but also connected to the ground at the service equipment. Sometimes the Code calls it the "identified conductor."

The term "ground wire" refers only to the wire from the service equipment to the ground. The Code calls it the "grounding electrode conductor." In the service equipment it is connected to the neutral wire of the incoming service wires. That wire is also connected or bonded to the metal cabinet of the service equipment, thus bonding them together and grounding them, as also the metal conduit or armor of armored cable that are anchored to the cabinet.

The term "ground*ing* wire" refers to a wire which does not carry current at all during normal operation; it is connected to parts of the installation such as frames of motors or clothes washers, the outlet boxes in which switches or receptacles are installed, and so on. The Code calls it the equipment grounding conductor. It runs with the current-carrying wires. In other words it is connected only to components that normally do not carry current, but which do carry current in case of damage to or defect in the wiring system, or the appliances connected to it. In the case of wiring with conduit, or cable with a metal armor, a ground*ing* wire as such is not installed separately, for the conduit or the armor of the cable serve as the ground*ing* conductor.

Green Wire: A wire used as a ground*ing* wire must be green, green with one or more yellow stripes, or bare. In this book it will be called just a green wire. It may never be used for any purpose except as the ground*ing* wire in a circuit.

The grounded wire may never be fused or protected by a circuit breaker, but every hot wire must be so protected. The grounded wire may never be interrupted by a switch; switches may be installed only in hot wires.

The grounded wire must run without interruption to all equipment operating at 120 volts, but not to anything operating at only 240 volts. Only hot wires run to 240-volt loads*. A separate ground*ing* wire runs to 240-volt loads (unless the conduit or the armor of armored cable serves as the ground*ing* conductor). If the wiring of 240-volt

---

* Anything that is connected to a circuit and consumes power constitutes a "load" on the circuit. The load might be a motor, a toaster, a lamp — anything consuming power. Switches do not *consume* power, therefore are not loads. A receptacle is not a load, but anything plugged into the receptacle is a load.

equipment is done using a 3-wire cable containing a white wire or bare wire, such white or bare wire may be used as the grounding conductor in a few cases that will be discussed as we go along. Most important of all: at the service equipment, all the grounding wires are connected to an equipment grounding busbar, which is bonded to the enclosure and connected to the neutral busbar (for all the white wires) which is in turn connected (grounded) to the water pipe and ground rod, by the ground wire that runs from the neutral busbar to the electrode system. The power supplier also grounds the neutral of the incoming service wires, at the transformer serving the building.

If the wire is *properly* grounded both at the transformer and at the building, it follows, that if you touch an exposed grounded wire at a terminal or splice, no harm follows, no shock, any more than if you touch a water pipe or a faucet, for the grounded wire and the piping are connected to each other. Any time you touch a pipe, you are in effect also touching the grounded wire.

Short Circuits and Ground Faults: If two hot wires touch each other at a point where both are bare, or if one touches a bare point in a ground*ed circuit* wire, it is called a *short circuit*. This rarely happens in the actual wires of a properly installed system, but often happens in cords to lamps or appliances, especially if badly worn or abused. When one hot wire at a bare point, for example where connected to a receptacle or switch, touches a grounded component, such as conduit, the armor of armored cable, or a ground*ing* wire, it is called a *ground fault*. The effect is the same: a fuse will blow or a breaker will trip.

How Grounding Promotes Safety: See Fig. 7-2 which shows a 120-volt motor with the grounded wire connected to the grounded neutral of your service equipment, and a fuse in the hot wire (actually the fuse and the ground connection would be at a considerable distance from the motor, at the service equipment cabinet, not close to the motor as shown, although there might be an additional fuse near the motor). If the fuse blows, the motor stops. You as the owner may inspect the motor, may accidentally touch one of the wires at the terminals of the motor. What happens? Nothing! The circuit is hot only up to the fuse. Between the fuse and the motor the wire is dead, just as if the wire had been cut at the fuse location. The other wire to the motor is grounded, so is harmless. You are protected. But do note that if the fuse is *not* blown and you touch the hot wire, you will receive a 120-volt shock, through your body, to the earth, and through the earth back to the neutral wire at the service equipment.

Now see Fig. 7-3 which shows the same circuit except with the fuse wrongly placed in the grounded wire instead of the hot wire. The motor will operate properly. The fuse blows, the motor stops. But the circuit is still hot, through the motor, up to the blown fuse. You touch one of the wires at the motor. What happens? You complete the circuit through your body, through the earth, to the neutral wire in your service equipment. You are directly connected across 120 volts, and as a minimum you will receive a shock, and at worst will be killed. The degree of shock and danger will depend on the surface on which you are standing, your general physical condition, and the condition of your skin at the contact. If you are on an absolutely dry surface you will note little shock; if you are on a damp surface as in a basement, you will experience a severe shock; if you are standing in water you will undergo maximum shock, or death.

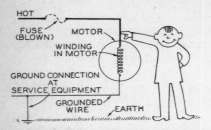

Fig. 7-2. A 120-v. motor properly installed except for a grounding wire.

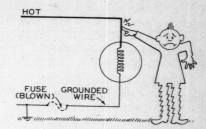

Fig. 7-3. A 120-v. motor installed with a fuse in the grounded wire. It is a dangerous installation.

Now see Fig. 7-4, which again shows the same 120-volt motor as in Fig. 7-2. But suppose the motor is defective, so that at the point marked G the winding inside the motor accidentally comes into electrical contact with the frame of the motor; the winding "grounds*" to the frame. That does not prevent the motor from operating. But suppose you choose to inspect the motor, touch just the frame of the motor. What happens? Depending on whether the internal ground between winding and frame is at a point nearest the grounded wire, or nearest the hot wire, you will receive a shock up to 120 volts, for you will be completing the circuit through your body back to the grounded wire. It is a potentially dangerous situation; shocks of a whole lot less than 120 volts can be fatal.

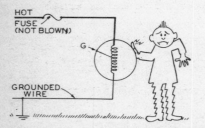

Fig. 7-4. The motor of Fig. 7-2, but the motor is defective. The winding is grounded to the frame. This is a dangerous installation.

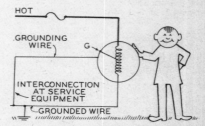

Fig. 7-5. The same as Fig. 7-4, but a grounding wire has been installed from the frame of the motor, to ground. It is a safe installation.

Now see Fig. 7-5 which shows the same motor, with the same accidental ground between winding (or cord) and frame, but protected by a grounding wire, connected to the frame of the motor, and running back to the ground connection at the service switch. When the internal ground occurs, current will flow over the grounding wire. It will

---

* Breakdowns in the internal insulation of a motor so that an electrical connection develops between the winding and the frame of the motor, are not uncommon. The entire frame of the motor becomes hot. The same situation arises if the motor is fed by a cord which becomes defective where it enters the junction box of the motor, so that one of the bare wires in the cord touches the frame. If there is no cord but the motor is fed by the circuit wires, a sloppy splice between the circuit wires and the wires in the junction box on the motor can lead to the same result: the frame of the motor becomes hot.

sometimes but not always blow the fuse. But even if the fuse does not blow, the grounding wire will protect you. The grounding wire reduces the voltage between the frame of the motor, to substantially zero as compared with the ground on which you are standing; you will not receive a shock *provided* that a really good job of grounding was done at the service. If there is a poor ground you will still receive a shock.

The ground*ing* wire from the frame of the motor (or from any other normally non-current-carrying component) may be green, or in many cases bare, uninsulated. If conduit, or cable with armor, is used, the conduit or armor serves as the grounding wire. Green wire may not be used for any purpose other than the ground*ing* wire. Other chapters will discuss just when a separate grounding wire must be installed.

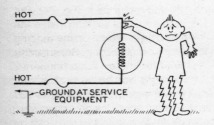

Fig. 7-6. A 240-v. motor properly installed except for a grounding wire. It is a safe installation only so long as the motor remains in perfect condition.

Fig. 7-7. The motor of Fig. 7-6, but the motor is defective. The winding is grounded to the frame. This is a dangerous installation.

Now refer to Fig. 7-6, which shows a 240-volt motor installed with a fuse in each hot wire (all hot wires must be fused, or protected by circuit breaker). Remember that in such 240-volt installations, the white grounded wire does not run to the motor, but is nevertheless grounded at the service equipment. If you touch both hot wires, you will be completing the circuit from one hot wire to the other, and you will receive a 240-volt shock. But if you touch only one of the wires, you will be completing the circuit through your body, through the earth, back to the grounded neutral in your service equipment, and you will receive a shock of only 120 volts: the same as touching the grounded wire and one of the hot wires of Fig. 7-1. The difference between 120-volt and 240-volt shocks may be the difference between life and death. (If you touch only the frame of the motor, you will not receive a shock.) This illustrates one of the benefits of proper grounding.

But assume that the motor becomes defective, that the winding of the motor becomes accidentally grounded to the frame, as shown in Fig. 7-7. You will recognize this as the same as Fig. 7-4, except that the motor is operating at 240 volts instead of 120 volts. Touching the frame will produce a 120-volt shock. But if the frame has been properly grounded as shown in Fig. 7-8, one of the fuses will probably blow, but even if it does not blow, touching the frame will not result in as severe a shock because the frame is grounded, again assuming that a really good ground was installed at the service.

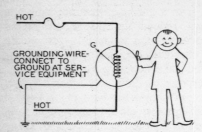

Fig. 7-8. The motor of Fig. 7-7, but a grounding wire has been installed from the frame of the motor, to ground. It is a safe installation.

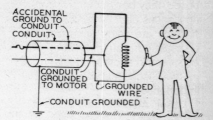

Fig. 7-9. A 120-v. motor properly installed, using wires in steel conduit or steel armor. The conduit is grounded at the service equipment cabinet; it is also grounded to the motor. It is a safe installation.

Advantages of Conduit or Armored Cable: Now let us consider a different aspect of the situation. In any of the illustrations of Figs. 7-2 to 7-8, if there is an accidental contact between any two wires of the circuit, this of course constitutes a short circuit and a fuse will blow, regardless of whether the short is between one of the hot wires and the grounded wire, or between the two hot wires. For all this to happen, there must be bare places on two different wires touching each other, which does not happen very often in a carefully installed job. But consider a wiring system in which all the wires are installed in a metal raceway: iron pipe called conduit, or using cable with a metal armor as in the cable commonly called "BX". The raceway or armor is grounded at the service equipment. It is also connected to the motor itself (assuming that the motor is not connected by a flexible cord and plug). No separate grounding wire is then required: if there is now an accidental ground from winding to frame, (or to the raceway or armor), it has a similar effect as a short between one of the hot wires and the grounded wire, because the grounded wire and the raceway or armor are connected to each other *and the ground* at the service equipment. A fuse will immediately blow, whether the motor operates at 120 or 240 volts. A considerable advantage has been gained, for accidental grounds that otherwise might remain undiscovered are automatically disclosed. All this is shown in Fig. 7-9.

In spite of the advantages of a metal raceway system, in some locations (especially farms) there are conditions that will be explained later, which make the use of a cable *without* metal armor more desirable. Then nonmetallic-sheathed cable is used, which in addition to the usual insulated wires has a bare uninsulated grounding wire which runs throughout the wiring system.

Continuous Grounds: As will be explained in Chap. 9, at every location where there is an electrical connection, an outlet box or switch box is used. When a grounded-neutral wiring system is used (and that is 100% of the time for installations of the kind discussed in this book), and you use metal conduit or cable with a metal armor, you must ground not only the neutral wire but also the conduit or cable armor. The neutral is grounded only at the service equipment (at the point where it is connected to the neutral of the service equipment), but the conduit or the armor must be securely connected to every box or cabinet. When lighting fixtures are installed on metal* outlet boxes, they become automatically grounded, through the conduit or armor. Motors or appliances directly connected to conduit or armor are automatically grounded.

If you are using nonmetallic-sheathed cable, it will contain an extra wire, a bare uninsulated ground*ing* wire, which must be carried from outlet to outlet, as will be explained in another chapter, providing a continuous ground. Regardless of the wiring method used, a *continuous* ground all the way back to the service, is of the utmost importance.

Other Advantages of Grounding: Suppose a 2400-volt line accidentally falls across your 120/240-volt service, during a storm. If the system is not grounded, you can easily be subject to 2400-volt shocks, and wiring and appliances will be ruined. If the system is *properly* grounded, the highest voltage of a shock will be much more than 240 volts, but very much less than 2400 volts.

Lightning striking on or even near a high voltage line can cause great damage to your wiring and your appliances, and can cause fire and injuries. Proper grounding throughout the system greatly reduces the danger.

Fig. 7-10. An ordinary receptacle and plug. Most receptacles are polarized, the wider slot being for connection to the grounded conductor. A polarized plug cannot be inserted into a non-polarized receptacle; replace the receptacle.

Fig. 7-11. A grounding receptacle, and plug. Plugs with either two or three blades will fit.

Receptacles: Many people are used to the original receptacle having two parallel openings for the plug, as shown in Fig. 7-10. Anything plugged into it duplicates the condition of Fig. 7-2. If the appliance is defective, and the owner handles it, he could receive a shock as shown in Fig. 7-4. This led to the development of what are called grounding receptacles, shown in Fig. 7-11. Note that such a receptacle has the usual two parallel slots for two blades of a plug, plus a third round or U-shaped opening for a third prong on the corresponding plug. In use the third prong of the plug is connected to a third or grounding wire in the cord, running to and connected to the frame of the motor or other appliance. (This third wire in the cord is green.) On the receptacle the round or U-shaped opening leads to a *green* terminal screw, which in turn is also connected to the metal mounting yoke of the receptacle. In turn, when installing the receptacle, a wire must be connected from this green terminal, to ground. If conduit or cable with armor is used, that wire runs to the outlet or switch box and the conduit or armor becomes the

* Nonmetallic boxes are also used; see chap. 17.

ground*ing* conductor. In the case of cable without armor, it will contain an extra wire, bare, uninsulated, in addition to the insulated wires, and this must be connected to the green terminal of the receptacle, and also grounded to the boxes. In this way, the frame of the motor or appliance is effectively grounded, leading to extra safety as shown and discussed in connection with Fig. 7-5. The details of how to connect the grounding wire from the green terminal will be discussed in Chap. 10.

If an appliance is connected by cord and plug, the Code in Sec. 250-45(c) requires a 3-wire cord and 3-prong plug on every refrigerator, freezer, air conditioner, clothes washer, clothes dryer, dishwasher, sump pump, aquarium equipment, and every hand-held motor-driven tool such as a drill, saw, sander, hedge trimmer and similar. It is not required on ordinary household appliances such as toasters, irons, radios, TV, razors, lamps and similar items. This might lead you to think you must install two kinds of receptacles, one for use with 2-prong plugs, and others for use with 3-prong plugs. Not so: the grounding receptacles are designed so that either plug will fit.

The Code requires that in all new construction, only grounding receptacles may be installed. If in an existing installation, you replace a defective receptacle the new one must be of the grounding type, if you can effectively ground it. If that is impossible, the replacement receptacle may *not* be the grounding type.

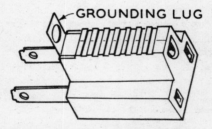

GROUNDING LUG

Fig. 7-12. This adapter permits a 3-prong grounding plug to be used with an ordinary receptacle.

What if you live in an older house with only the non-grounding receptacles, and you want to use an appliance with a 3-wire cord and a 3-prong plug, which of course will not fit the older receptacles? Use a "two-to-three-wire" adapter shown in Fig. 7-12. Note that it has a *green* terminal lug on its side. Remove the screw that holds the face plate to the receptacle, plug in the adapter, and reinstall the screw through the green lug, which is then in contact with the mounting yoke of the receptacle. This procedure is by no means the complete equivalent of having grounding receptacles in your home, but it does permit you to use appliances with 3-prong plugs, and does provide some degree of protection, *if* the receptacle yoke is grounded to the grounded metal box. But, if your home is wired using metal conduit or armored cable, it would be far better to replace the 2-wire receptacle with one of the grounding type, of course adding a grounding wire from the green terminal of the receptacle to the grounded metal switch box, as will be discussed in Chap. 10.

How Dangerous are Shocks? Most people think it is a high voltage that causes fatal shocks. This is not necessarily so. The amount of current flowing through the body determines the effect of a shock. A milliampere is 1/1000th of an ampere. A current of one milliampere through the body is just barely perceptible. One to eight milliamperes causes mild to strong surprise. Currents from eight to 15 milliamperes are unpleasant, but usually the victim is able to free himself, to let go of the object that is causing the shock. Currents over 15 milliamperes are likely to lead to "muscular freeze" which prevents the victim from letting go and often lead to death. Currents over 75 milliamperes are almost always fatal; much depends on the individual involved.

Of course the higher the voltage, the higher the number of milliamperes that would flow through the body, under any given set of circumstances. A shock from a relatively high voltage while the victim is standing on a completely dry surface will result in fewer milliamperes, than a shock from a much lower voltage while he is standing in water. Many deaths have been caused by shock on circuits considerably below 120 volt; many have survived shock from circuits of 600 volts and more. How to help the victim of a shock will be discussed in Chap. 19.

GFCI: The use of a grounding wire in a 3-wire cord with a 3-prong plug and a grounding receptacle reduces the danger of a shock when using, for example, a portable tool, but it does not eliminate the danger 100%. Cords can be defective or wrongly connected; millions of tools with 2-wire cords are still in use. Some people foolishly cut off the grounding prong on a 3-prong plug, because they have only 2-wire receptacles (see Figs. 7-10 and 7-11).

Under normal conditions the current in the hot wire and the grounded wire are absolutely identical. But if the wiring, or a tool or appliance, is defective and allows some current to leak to ground, then a ground fault circuit interrupter (abbreviated GFCI) will sense the difference in current in the two wires. If the fault current exceeds the trip level of the GFCI, which is between 4 and 6 milliamperes (thousandths of an ampere), the GFCI will disconnect the circuit in as little as 1/40 second.

The fault current, much too low to trip a normal breaker or blow a fuse, could possibly flow through a person in contact with the faulty equipment and a grounded surface. The use of a GFCI is a most desirable safety precaution, especially when using electrical equipment outdoors, for standing on the ground (especially if wet) greatly increases the likelihood and especially the severity of a shock. The GFCI you install must be rated in amperes and volts to match the rating of the outlet or circuit it is to protect.

The GFCI is not an inexpensive device, but should be considered insurance against dangerous shocks. It is not to be considered as a substitute for grounding, but as supplementary protection which senses leakage currents too small to operate ordinary branch circuit fuses or circuit breakers. The GFCI will not prevent a person who is part of a ground-fault circuit from receiving a shock, but it will open the circuit so quickly that the shock will be below levels which will inhibit breathing or heart action, or the ability to "let go" of the circuit.

The Code in Sec. 210-8 requires a GFCI on all 15- and 20-amp. residential outdoor, bathroom and some garage receptacles, and for temporary receptacles used during building construction. Swimming pool installations, not covered in a book of this size, also require GFCI protection. See Article 680 of the Code.

Three types of GFCI are available. The *first* or separately-enclosed type is available for 120-volt 2-wire, and 120/240-volt 3-wire circuits up to 30 amp. It is most often used in swimming pool wiring, installed at any convenient point in the circuit. The *second* type combines a 15, 20, 25 or 30-amp. circuit breaker and a GFCI in the same plastic case, is installed in place of an ordinary breaker in your panelboard, and is available in 120-volt 2-wire, or 120/240-volt 3-wire types (which may also be used to protect a 2-wire 240-volt circuit). It provides protection against ground faults *and overloads* for all the outlets on the circuit (at any time you can replace an ordinary breaker in your panelboard with one of these combination breakers). Each GFCI circuit breaker has a white pigtail which you must connect to the grounded (neutral) busbar of your panelboard. You must connect the white (grounded) wire of the circuit to a terminal provided for it on the breaker. The *third* type, a receptacle and a GFCI in the same housing, provides only ground-fault protection to the equipment plugged into that receptacle; or if of the "feed-through" type, to equipment plugged in to other ordinary receptacles installed "downstream" on the same circuit. This type is most convenient in replacing existing receptacles where GFCI protection is desired or required.

Regardless of the type or brand of GFCI you install, it is most important that you most carefully follow the installation and periodic testing instructions that come with it. Every GFCI has a test button for easy verification of its functional operation.

Even where wiring, tools, and appliances are in perfect condition and there is no ground fault, be on the lookout for these installation problems which will cause tripping of a GFCI: a 2-wire type (other than an end-of-run GFCI receptacle) is connected in a 3-wire circuit; the white circuit conductor is grounded on the load side of the GFCI; or, the protected portion of the circuit is excessively long (250 ft., as a rule of thumb. Longer circuits may develop a capacitive leakage to ground). Of course, the GFCI is *designed* to trip if the cords or tools supplied are in poor repair and provide a path for current to leak to ground.

Remember: the GFCI will not prevent shock, but will render shocks relatively harmless; and, it will not protect a person against contact with both conductors of the circuit at the same time, unless there is also a current path to ground. The GFCI may be pictured as an electronic adding machine which constantly monitors the current out to a load and back again. The GFCI acts to quickly disconnect the circuit only when the current out to the load and the current returning differ by 5/1000 amp. or more.

Chapter 8

THE SERVICE ENTRANCE

The service entrance consists of the following components:

1. The insulators on the outside of the building, at which point the power supplier's wires end.
2. The meter socket.
3. Cable (or wires in conduit) from the insulators to the meter socket, and then into the building.
4. The disconnecting means and overcurrent protection (circuit breakers or fuses).
5. The ground connection.

This chapter will discuss the design and installation of a service entrance in a city home. Installations on farms are somewhat different and will be discussed in Chap. 17.

2-wire Entrances: When only two wires are installed, only 120-volt current is available. One of the wires is white (sometimes bare) beginning at the insulators, and is grounded. It is *not* a neutral wire; there cannot be a neutral in a 2-wire circuit. The hot wire is black.

3-wire Entrances: When three wires are installed, both 120 and 240 volts are available. This was shown in Fig. 7-1 in the preceding chapter. The voltage between N and A, or between N and B, is 120 volts; between A and B it is 240 volts. The wire N is white (sometimes bare) beginning at the insulators. It is grounded, and up to the service equipment inside the building, it is a *neutral* wire. The two hot wires can be any color except white or green, but are usually black and red, or both black. Note that even if no 240-volt appliances were used, the 3-wire system has advantages (other than providing two separate voltages) which will be discussed in Chapter 13. In any event, the 3-wire service is entirely standard today.

Ratings of Service Switch or Breaker: Service switches are rated at 30, 60, 100 or 200 amp. Service breakers are rated at 30, 40, 50, 60, 70, 90, 100, 125, 150, 175 and 200 amp. Equipment larger than 200 amp. is of course available for larger installations.

Multiply the amperage rating by the voltage (240) to determine the maximum wattage available with any size. For example a 60-amp. switch or breaker will make available 60 × 240 or 14 400 watts; a 100-amp. size will make available 100 × 240 or 24 000 watts, and so on.

In wiring a residence, the service equipment must have a rating of 100 amp. or higher, unless there are five or fewer 2-wire circuits, which rarely is the case. At one time, smaller equipment was acceptable; the present requirement merely reflects the fact that such installations after a few years turned out to be too small.

In wiring other than residences, a 30-amp. service may be installed if there are not over two 2-wire circuits. In all other cases, the equipment must be adequate for the load involved, but never less than 60 amp. See Code Sec. 230-79.

Determining Size Service Required: Determine the probable load as follows:

Area of house* _____sq. ft. × 3 watts . . . . _____watts
Two appliance circuits . . . . 3000 watts†
One laundry circuit . . . . 1500 watts
Range if used. . . . . _____watts‡
Water heater if used . . . . _____watts
Other heavy permanently-connected appliances. . . . . _____watts
                                                    Total . . . . _____watts

If the total above is under 10 000 watts, *and* if there are to be no more than five 2-wire circuits, a 60 amp. service is permitted, but not recommended by the author because it will soon be found to be too small. If the total above is over 10 000 watts (as it will be if a water heater or range is installed) a 100-amp. is the smallest permitted by Code. Refer once more to the table in Chap. 3, showing watts consumed by appliances, and you will agree that a 60-amp. service is too small.

The 100-amp. service provides a maximum capacity of 24 000 watts. If the total of the "net computed watts" above comes to more than 24 000 watts, does it follow that a service larger than 100-amp. is necessarily required? Not at all. Not all of the lights and appliances listed in the tabulation will be in use *at the same time.* But do remember that the assortment of appliances you include in your first tabulation represent your *present* ideas; as time goes on, you will add more.

A 150- or 200-amp. service is desirable for a modern home, and a "must" for a really large home. In some areas, local Codes are already requiring services larger than 100 amp. The larger service costs relatively little more than one of the 100-amp. size.

Size of Service Entrance Wires: The Code in Sec. 230-41 permits No. 8 to be used if there are not over two 2-wire circuits, and No. 6 if there are not over five 2-wire circuits. But homes always have more than five circuits, so No. 8 and No. 6 may be used only in installations other than homes. Since the 100-amp. service is the minimum for homes, select a wire with ampacity of 100 or more amperes, from the table in Chap. 4.

Circuit Breakers or Fused Equipment? Inside the house you must provide equipment so that you can disconnect all the wiring from the power source, and you must provide either circuit breakers or fuses to protect the installation as a whole, as well as each branch circuit individually. Most installations today use circuit breakers because of their many advantages. If one trips, restore service by merely flipping its handle (after correcting the problem that created the overload that tripped the breaker). You can also disconnect one circuit by using the breaker as you would a switch. The breakers will carry nondangerous temporary overloads that would blow ordinary fuses.

---

* Figured as outlined in Chap. 5.

† Each of these circuits will provide 20 × 120 or 2400 watts, but for the purpose of this calculation, you may use only 1500 watts per circuit, 3000 watts altogether, because it is not likely that both circuits will be loaded to capacity at the same time.

‡ Per Code, you need to allow only 8000 watts if the rating of your range (or counter top plus separate oven) is 12000 watts or less. If the rating is over 12000 watts, add 400 watts for each additional kilowatt or fraction thereof.

You will need a single-pole breaker for each 120-volt circuit, a 2-pole (double-pole) breaker for each 240-volt circuit. A 2-pole breaker occupies the same space in a cabinet, as two single-pole breakers.

Solid Neutral: Whether your service equipment contains breakers or fuses, the incoming neutral wire may never be interrupted by a breaker or fuse. Therefore in a 3-wire 120/240-volt installation, while there are three service-entrance wires, the main breaker is only a 2-pole, and in fused equipment there are only two main fuses. The neutral is called a "solid neutral" — a neutral not protected by breaker or fuse. Such equipment then is called "3-wire SN."

Neutral and Grounding Busbars: The cabinets of service equipment will contain a copper grounding busbar, on which are installed several large solderless connectors for the heavy incoming neutral wire and the ground wire, plus as many smaller connectors or terminal screws as needed, one for the grounded wire of each 120-volt circuit. If a wiring method such as Type NM Cable is used, with separate equipment grounding wires, a separate grounding busbar will have to be installed, as the neutral busbar will not likely have enough terminals for each white wire plus each grounding wire. As purchased the neutral busbar may be insulated from the cabinet, and if you use this equipment as *service* equipment (but *not* if used on the load side or "downstream" from the service equipment), you must bond it to the cabinet. This is a very simple matter. In some brands of equipment, the neutral busbar is provided with a heavy bonding screw that you must tighten securely. In other brands the cabinet will contain a flexible metal strap that is already bonded to the cabinet, and you must connect its free end to one of those connectors or screws on the neutral busbar. The equipment grounding busbar will always be bonded to the enclosure, whether within the service equipment enclosure or at a remote panelboard.

Selecting Breaker Equipment: Most service-entrance equipment as purchased will contain *one* large (100-amp. or larger) 2-pole main breaker that will protect the entire installation, and will let you disconnect the entire installation from the power source by merely turning this one breaker to the *off* position. But there must be more breakers to protect individual circuits: single-pole for 120-volt circuits, and 2-pole for 240-volt circuits. These branch-circuit breakers are not part of the equipment as purchased, but the cabinet contains an arrangement of busbars into which you can plug individual breakers, single-pole or 2-pole as you wish, and as many as there is room for. Such equipment and its internal wiring diagram are shown in Fig. 8-1.

Somewhat less popular (and in a few localities prohibited by local ordinance) are what are called "split-bus" service-entrance panelboards. The Code permits two *main* disconnects in one service panelboard. To disconnect everything from the power source it is necessary to turn *both* of them to the *off* position. In addition to the main breakers, the cabinet may contain as many breakers as you wish (up to the capacity of the panelboard), single-pole or two-pole, each protecting its own circuit, *but* all of these smaller breakers must be protected by one of the main breakers. See the wiring diagram of Fig. 8-2. An alternate arrangement would be for one of the main breakers to protect all of the branch circuits in the service panelboard, and the other main to supply feeder to a panelboard located elsewhere.

In a few areas it is common practice to use an outdoor cabinet containing the meter socket and the service disconnects. See Fig. 8-3, showing a meter socket and single main. In some mild climate areas all of the branch circuits could also originate in this enclosure.

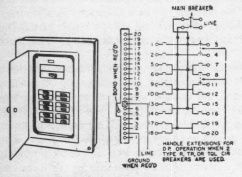

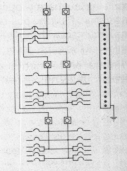

Fig. 8-1. A circuit-breaker cabinet, with main breaker and space for 20 circuits.

Fig. 8-2. Diagram for a split-bus circuit-breaker cabinet.

Fig. 8-3. A combination of meter socket plus circuit breaker in the same cabinet is very popular in some localities.

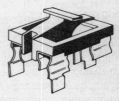

Fig. 8-4. In fused service equipment, cartridge fuses are usually installed on pullout blocks. Replace fuses while the block is in your hand.

Selecting Fused Equipment: In very old houses you will find service equipment that consists of a main switch with two hinged blades and an external handle, used to turn the entire load in the building on and off. The switch cabinet also contains two *main* fuses. Usually it also contains as many smaller fuses as needed to protect all the individual branch circuits in the house; sometimes these fuses are in a separate cabinet.

Instead of a switch with an external handle, the type widely used in residential work has two main fuses mounted on a pull-out block as shown in Fig. 8-4. Insert the fuses into their clips while the pull-out is in your hand. One side of each fuse clip has long prongs, so that the entire pull-out has four such prongs. The equipment in the cabinet contains four narrow open slots, but no exposed live parts; the live parts are behind the insulation. Plug the pull-out with its fuses into these slots; the four prongs on the pull-out make contact with the live parts, thus completing the circuit. Plugging the pull-out into its holder is the same as closing a switch with hinged blades; removing it is the same as opening such a switch. If you wish, you can insert the pull-out upside-down, which leaves the power turned off.

Fig. 8-5. A 100-amp. switch. Besides main fuses it has plug fuses for twelve 120-volt circuits, and cartridge fuses for two 240-volt circuits.

Just as in the case of circuit breaker equipment, such service equipment usually contains one main pull-out with large fuses, protecting the entire load, plus additional pull-outs to protect 240-volt circuits with large loads, such as ranges, etc., plus as many fuseholders for *plug* fuses as are needed to protect individual branch circuits: one for each 120-volt circuit, two for each 240-volt circuit. A service panelboard of this type is shown in Fig. 8-5. The wiring diagram is the same as shown in Fig. 8-1, except that fuses are used in place of breakers.

And again, as in the case of breaker equipment, the fused equipment *may* contain two *main* pull-outs, connected internally so that the service wires will run directly to each. To disconnect the load completely, you must pull out both of the main pull-outs. And as in the case of breaker equipment, the branch-circuit fuses must be protected by one of the main fused pull-outs. The wiring diagram of connections within the equipment again is that of Fig. 8-2 except that fuses replace the breakers.

A word of explanation regarding fused switches is in order. If the fuses are of the cartridge type, the "60-amp." holder will accept fuses rated from 35- to 60-amp. The "100-amp." holder will accept fuses rated from 70- to 100-amp. The "200-amp." holder will accept fuses rated from 110- to 200-amp. Holders for plug fuses of course will accept fuses up to 30 amp. In new work, Type S adapters must be used, so you can always install only the proper fuse to match the ampacity of the wires you are using.

How Many Circuits? This subject has already been discussed in Chapter 6 and need not be repeated here. Select equipment containing the required number of branch-circuit fuses or breakers, but be wise: add a few spares for future use. You cannot today foresee all the additional electrical equipment that you will use at some future time.

Location of Service Equipment: The Code requires that this equipment be located as close as practicable to the point where the wires enter the house. In other words, the service wires must not run 10 or 20 ft. inside the house before reaching this equipment. So, decide where to put your switch or breakers before deciding where the wires are to enter the house.

The heaviest loads in a house, the loads that will consume the largest amperages, are located in kitchen and basement. Therefore, locate the service equipment so that the circuits using large wires (to range, clothes dryer, and similar equipment) will be as short as possible. A point in the basement more or less under the kitchen is desirable, if the service wires can be brought into that location directly from outside. Some prefer to locate the equipment in the kitchen, where it is convenient for replacement of fuses or resetting of breakers.

Location of Meter: The outdoor type of meter is used in nearly every case today. The meter itself is weatherproof and in turn is plugged in to a weatherproof socket, both of which are shown in Fig. 8-6. The socket is usually, but not always, supplied by the power supplier but installed by the contractor, usually about 5 ft. above the ground. In the case of indoor equipment, the meter is installed on a substantial board near the service equipment.

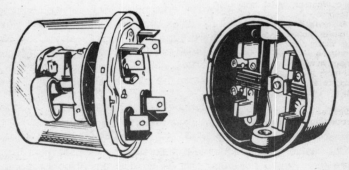

Fig. 8-6. A typical outdoor weatherproof kilowatthour meter and its socket.

Installation of Service Entrance: Up to now we have discussed the major parts that make up the service entrance, and how to select the right kind and size. All these parts must now be installed. Fig. 8-7 shows a typical entrance. You will have to install insulators on the outside of the building and the power supplier will run his wires up to that point. You must provide wires from the end of the power supplier's wires, down to the meter socket, and then on into the house to the service equipment. You can use special cable for that purpose, or run wires through conduit. After installing the service equipment, you must properly ground the installation. Then you are ready to wire the branch circuits.

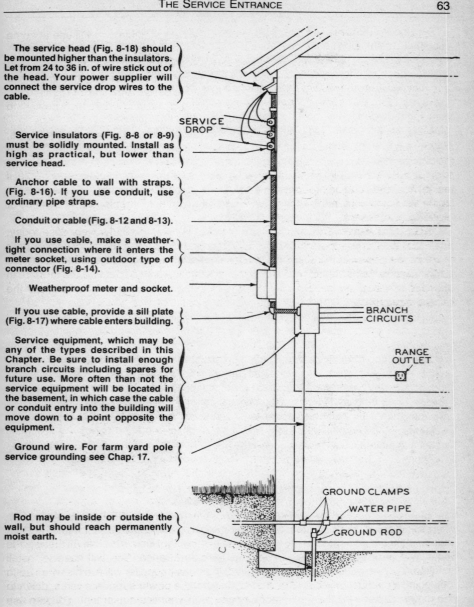

The service head (Fig. 8-18) should be mounted higher than the insulators. Let from 24 to 36 in. of wire stick out of the head. Your power supplier will connect the service drop wires to the cable.

Service insulators (Fig. 8-8 or 8-9) must be solidly mounted. Install as high as practical, but lower than service head.

Anchor cable to wall with straps. (Fig. 8-16). If you use conduit, use ordinary pipe straps.

Conduit or cable (Fig. 8-12 and 8-13).

If you use cable, make a weather-tight connection where it enters the meter socket, using outdoor type of connector (Fig. 8-14).

Weatherproof meter and socket.

If you use cable, provide a sill plate (Fig. 8-17) where cable enters building.

Service equipment, which may be any of the types described in this Chapter. Be sure to install enough branch circuits including spares for future use. More often than not the service equipment will be located in the basement, in which case the cable or conduit entry into the building will move down to a point opposite the equipment.

Ground wire. For farm yard pole service grounding see Chap. 17.

Rod may be inside or outside the wall, but should reach permanently moist earth.

SERVICE DROP

BRANCH CIRCUITS

RANGE OUTLET

GROUND CLAMPS

WATER PIPE

GROUND ROD

Fig. 8-7. A cross-sectional view of a typical service entrance.

Service Insulators: The incoming service wires are anchored on service insulators installed on the building. They should be installed a foot or so *below* the highest practical point, so that the wires to the inside of the house, where they emerge from the service head on the conduit or cable, will slope downward, towards the insulators, to prevent water from following the service wires into the service head. This should be clear from Fig. 8-7. In most localities racks of the type shown in Fig. 8-8 are used; the spacing between insulators is usually 8 in. although in some localities those with 6 in. spacing are acceptable. In place of multiple racks, individual screw-point insulators o Fig. 8-9 are often used.

Fig. 8-8. Insulator rack for supporting three wires. Be sure to anchor rigidly to building.

Fig. 8-9. Screwpoint insulator.

Fig. 8-10. This insulator clamps to pipe support.

The Code in Sec. 230-24(b) requires that cabled service wires not over 150 V to ground, or drip loops, be kept 10 ft. above finished grade at the point of attachment to the building; 12 ft. above residential property and driveways, and commercial property not subject to truck traffic; 18 ft. over parking areas and driveways other than residential, and farm properties, subject to truck traffic. No wires may come closer than 3 ft. to a window*, door, porch, fire escape or similar location from which they might be touched. If they pass over all or part of a roof that has a rise of 4 in. or more per foot, they must be kept at least 36 in. from the nearest point. If the roof is flatter (less than 4 in. o rise per foot) they must be kept at least 8 ft. from the nearest part of the roof.

Masts: Many houses built today are of the rambler or ranch-house type, where it is difficult to maintain such clearances if the insulators are mounted directly on the side o the house. In that case use a mast, of which several types are on the market; typica construction is shown in Fig. 8-11. In such installations the service conduit extends upward through the overhang of the roof; the conduit becomes the support fo insulators of the type shown in Fig. 8-10. Unless your power supplier has othe requirements, the service conduit should not be smaller than 2-in. rigid metal conduit Smaller conduit could be used if braced or guyed for extra strain support, or could be attached to a 4 x 4 in. timber, as could service entrance cable. If the roof overhang does not exceed 48 in., the service wires may be as low as 18 in. above the roof.

Service-Entrance Cable: Service-entrance cable is usually used to bring wires into the building. It is illustrated in Fig. 8-12. One of the wires is not insulated and consists o a number of fine wires wrapped around the insulated wires. In use, the small bare wires are twisted together to make one larger wire, as shown in Fig. 8-13. The bare wire may

*There is no restriction to wires above *the top* of a window.

be used only for the grounded neutral. Over all comes a fabric braid or outer protective layer. This usually has a painted finish which may be painted to match the building. As already explained, 30- and 60-amp. switches are not permitted in residential work, but may be used in other buildings. Use No. 8 wire with a 30-amp. switch, No. 6 with a 60-amp. With a 100-amp. switch, use cable with No. 2 wires, and with a 200-amp. use cable with No. 3/0 wires.

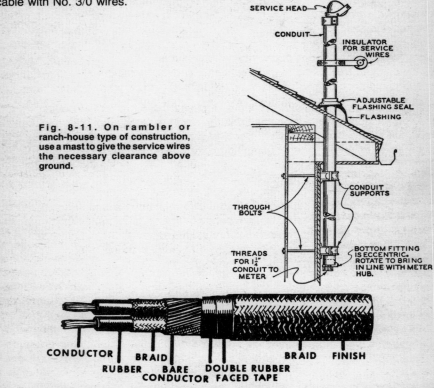

Fig. 8-11. On rambler or ranch-house type of construction, use a mast to give the service wires the necessary clearance above ground.

Fig. 8-12. In service entrance cable the neutral wire is not insulated, but is bare, and wrapped spirally around the insulated wires. The picture shows 3-wire cable.

Fig. 8-13. When using the cable, the separate strands of the neutral bare wire are gathered into a bunch making one large wire.

STRANDS OF BARE CONDUCTOR, TWISTED

If the installation served by the service equipment has no significant loads operating at 240 volts, all wires in the cable, including the neutral, must be the same size. But if the installation has an electric range or water heater (or other 240-volt loads) consuming about one third of the total watts, then you may use cable with the neutral one size smaller than the insulated wires: for example, No. 6 insulated wires with a No. 8 bare neutral.

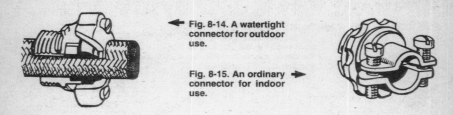

◄ **Fig. 8-14. A watertight connector for outdoor use.**

Fig. 8-15. An ordinary connector for indoor use. ➤

Wherever the cable enters a service equipment cabinet or a meter socket, it must be securely anchored with a connector. Out-of-doors a weatherproof connector must be used; a common type is shown in Fig. 8-14. The connector consists of a body and a heavy block of rubber, and a clamping nut or cover which compresses the rubber against the cable, making a water-tight joint. Indoors a less expensive connector of the type shown in Fig. 8-15 may be used instead. Cable is anchored to the building with straps, one type of which is shown in Fig. 8-16.

At the point where the cable enters the building, you must take steps to prevent rain water from following the cable into the building. A waterproof putty-like compound is suitable; pack it tightly into all openings around the cable. A sill plate as shown in Fig. 8-17 is convenient but not required by the Code. Waterproofing compound is usually supplied with the plate.

◄ **Fig. 8-16. Entrance cable must be anchored to the building every 4½ ft. Use straps of the type shown.**

Fig. 8-17. Use a sill plate packed with sealing compound, where cable enters building. It keeps water out. ➤

When you install service entrance cable, cut a length long enough to reach from the meter socket to a point at least a foot above the topmost insulator, plus another two or three feet. On this last additional length, remove the outer braid over the spirally-wrapped neutral wire. Unwind these bare wires from around the cable, and twist them into a single wire as shown in Fig. 8-13. Then install a service head of the general type shown in Fig. 8-18, letting the individual wires project through separate holes in the insulating block in the service head. The service head is designed to prevent water from entering the top of the cable.

Anchor the service head on the building about 12 to 18 in. above the topmost insulator, so that after the connection has been made to the power supplier's wires, rain will tend to flow away from the service head rather than into it. If it is utterly impossible to locate the service head above the insulators, be sure to provide drip loops as shown in Fig. 8-19; this is another way of keeping the rain out of the cable. Your power supplier will connect the ends of the wires in the cable to the service-drop wires.

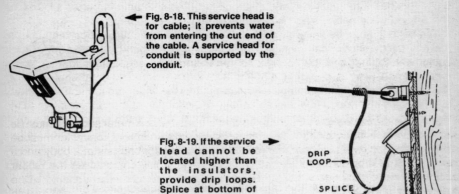

Fig. 8-18. This service head is for cable; it prevents water from entering the cut end of the cable. A service head for conduit is supported by the conduit.

Fig. 8-19. If the service head cannot be located higher than the insulators, provide drip loops. Splice at bottom of loop, and insulate.

DRIP LOOP

SPLICE

CABLE

Anchor the bottom end of the cable to the meter socket, and connect the three wires to the terminals in the socket, as shown in Fig. 8-20, which also shows the next length of cable running to the inside of the building. The cable must be supported on the building within 12 in. of the service head and meter socket, and additionally every 4½ ft. Straps such as shown in Fig. 8-16 are suitable.

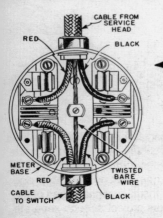

CABLE FROM SERVICE HEAD
RED
BLACK
METER BASE
RED
CABLE TO SWITCH
TWISTED BARE WIRE
BLACK

Fig. 8-20. How wires are connected to the meter base or socket. The bare, uninsulated wire is always connected to the neutral center contact of the socket.

Fig. 8-21. When entrance wires run through conduit, use entrance cap of this type, and special entrance ell with removable cover, at the bottom.

Service Entrance with Conduit: Instead of service entrance cable, separate Type TW, THW or RHW (but not Type T) wires inside conduit may be used. The neutral wire is always white; in some localities bare, uninsulated wire is used. The other two wires are black, or black and red. To determine the size conduit required, see table in Chap. 12.

Conduit is available in three types: rigid, intermediate, and thin-wall, all described in Chap. 12. Cut pieces as long as required, prepare them in the manner outlined in Chap. 12. Install a service head at the top, of the type shown in Fig. 8-21. At the point where the conduit is to enter the house, use an entrance ell, one type of which is shown in Fig. 8-21. This has a removable cover which makes it easy to help the wires around the sharp bend while pulling them into the conduit.

The conduit must be securely anchored to the service equipment cabinet, both to make a good mechanical joint, and to provide a good bond between the service equipment cabinet and the conduit, for grounding purposes. Use a locknut and a grounding bushing as shown in A of Fig. 8-25.

After the conduit itself is completely installed, pull the wires into it. For the relatively short lengths involved, the wires can usually be pushed in at the top and down to the meter socket; from there, other lengths are pushed through to the inside of the house. If the conduit is quite long or has bends in it, use fish tape; a length of galvanized clothes line or similar wire will serve the purpose for short lengths. Push the fish tape into the conduit, tie the electrical wires to it, and pull them in. The use of fish tape is more fully explained in Chap. 12.

Concentric Knockouts: The cabinets of service switches or circuit breakers are too small to permit knockouts of all the sizes that might be required in all circumstances. Therefore "concentric" knockouts are provided, as shown in Fig. 8-22. Remove the center section if you need the smallest size; remove the two smallest sections if you want the next size, and so on. You must be very careful, lest you remove more of this intricate knockout than is required for your particular installation.

Fig. 8-22. Concentric knockouts are convenient, but great care must be used in removing parts of them.

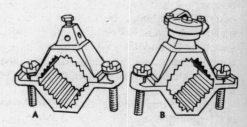

Fig. 8-23. Typical ground clamps for connecting ground wire to water pipe.

Connections at Service Equipment: The switch or circuit breaker has two heavy terminals to which the two incoming hot wires are connected. It also has a neutral busbar, on which you will find several large connectors, and a number of smaller terminals. Connect the incoming white or bare neutral to one of the large connectors, and the ground wire to the other. The smaller terminals are for the white wires of the individual circuits or feeders. If there are enough terminals, bare grounding wires of circuits can also be connected to the neutral bar *in service equipment.* If additional terminals are required, an accessory grounding bar must be used. The enclosure will have mounting means for this ground bar.

Grounding: Every installation of the type discussed in this book must be grounded, and the importance of doing a good job can't be overemphasized. Grounding is accomplished by running a ground wire from the incoming neutral service wire (from a point where it is connected to the neutral busbar in the service equipment) (a) to a grounding electrode system, or where not available (b) to a grounding electrode. Where available at each building or structure, these electrodes must be bonded together to form a grounding electrode system: 10 ft. or more of buried metal water pipe; the effectively grounded metal frame of the building; a concrete-encased electrode (20 ft. or more of ½ in. reinforcing steel, or No. 4 copper wire, encased by at least 2 in. of concrete and located near the bottom of a concrete foundation or footing that is in direct contact with the earth); or a ring of No. 2 or larger bare copper encircling the building. A metal underground water pipe must *always* be supplemented by one additional electrode. Should none of the above be available, uncoated metal underground gas piping (where acceptable to the gas supplier *and* to the local inspector); other underground metal piping or tanks; an 8 ft. driven ½ in. copper or ⅝ in. steel rod or ¾ in. galvanized steel pipe; or a 2 sq. ft. buried metal plate may serve as the grounding electrode. In any case, the armor of armored cable, or the conduit of a metal conduit system, become grounded because they are anchored to the cabinet of the service equipment.

Important: The neutral busbar of the service equipment as purchased may be insulated from the cabinet, but is (a) equipped with a special screw that you must *firmly* tighten to bond it to the cabinet, or (b) the cabinet has a flexible metal strap solidly bonded to it, and you must connect the opposite end of it to the grounding busbar. The neutral busbar *must* be bonded to the cabinet if the switch is used as service equipment; it must *not* be bonded if the switch is used for example as a disconnecting switch for an appliance or other load, or used as a panelboard on the load side of a service. The technical reasons for grounding were discussed in Chap. 7, and a good, carefully installed ground *is absolutely necessary* if the completed installation is to be safe.

Ground Wire: There is no objection to using insulated wire, but bare wire costs less and is generally used for the ground wire. If your service has No. 2 or smaller wires, you may use No. 8 wire for the ground, but it must be enclosed in armor or conduit. You will find it much less costly and far more practical to use No. 6 which does not have to be protected like No. 8.

If your service has No. 1 or No. 1/0 wire, No. 6 is the minimum size permitted for the ground, and if it has No. 2/0 or 3/0, No. 4 is the minimum. Neither size needs to be protected by armor or conduit, but must be fastened to the surface over which it runs; use staples for the purpose. Naturally it must be installed in a way that minimizes danger of physical damage after installation. In any event keep the ground wire as short as possible.

At the service equipment, connect the ground wire to the neutral busbar to which you have connected the neutral of the service. Run it to the grounding electrode(s) and there connect it to a ground clamp of the general style shown in *A* of Fig. 8-23. This clamp must be made of galvanized iron if installed on an iron pipe, but copper or bronze if it runs to a copper pipe or rod.

Bonding at Service: If your service-entrance wires are in conduit, follow *A* of Fig. 8-25. Install a grounding bushing (shown in Fig. 8-26) on the end of the conduit inside the cabinet; run a jumper (bare or insulated, of the same size as the ground wire) from the bushing to the grounding busbar. The screw of the bushing bites down into the metal of the cabinet, thus contributing toward a good continuous ground; it also prevents the bushing from turning and loosening which would impair the continuity of the ground.

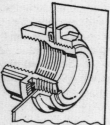

Fig. 8-24. Two-piece threaded hub. Serrations on both pieces bite into enclosure metal, making a bonding jumper unnecessary. O-ring in outside piece forms a raintight seal.

When you use a threaded hub bolted to the enclosure, or of the two-piece type shown in Fig. 8-24, bonding of the conduit to the enclosure is automatic and the grounding bushing and jumper are not necessary.

If your service uses service-entrance cable with its bare neutral wrapped around the insulated wires, as was shown in Fig. 8-12, the grounding bushing is not required. Follow *B* of Fig. 8-25.

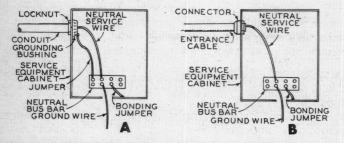

Fig. 8-25. Typical grounding connections. Only the white wire is shown. The wire marked "bonding jumper" is usually part of the service equipment as purchased; it could be either a screw, or a flexible strap.

Fig. 8-26. A grounding bushing.

Some prefer to install the ground wire enclosed in armor similar to that of armored cable, or in conduit. If you use armored ground wire, install a connector on the cable at the cabinet, and use a ground clamp of the type shown at *B* of Fig. 8-23; connect the wire to the terminal screw, and clamp the armor in the separate clamping device of the ground clamp.

If the ground wire is protected by conduit, anchor the conduit to the cabinet with a locknut and bushing, preferably a grounding bushing. A jumper from that bushing to the neutral busbar is required, as shown for the service conduit in *A* of Fig. 8-25. Use a ground clamp into which you can thread the conduit, and connect the ground wire to a terminal screw of the clamp.

It is of the utmost importance that when using either armor or conduit to protect the ground wire, the armor or conduit be securely and permanently bonded at *both* ends, to the ground wire and to the cabinet and ground clamp. Unless this is done, the resultant ground will be *very* much less effective than when using unprotected wire. The explanation of this fact is far beyond the scope of this book.

Only *one* Ground Wire: There is never more than one ground wire from the service equipment to the grounding electrode(s). If grounding of any equipment or enclosures is required, the ground*ing* conductor must run back to the service equipment, where it is bonded to the enclosure, the neutral, and the ground wire. Metal conduit, or the metal armor of armored cable, serves as the grounding conductor, eliminating the need for a separate grounding conductor. For grounding at separate buildings, as on farms, see Chap. 17.

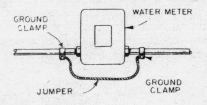

Fig. 8-27. Install a jumper around the water meter.

Water Meters: A jumper must be installed around the water meter as shown in Fig. 8-27. Use two ground clamps and a length of wire of the size used for the ground. This prevents the ground from being made ineffective if the water meter is removed; some water meters also have insulating joints.

Made Electrodes: If there is no underground water system, a good ground becomes a problem. Proper method of installation is shown in Chap. 17. But do note that if only a ground rod is used, the ground wire never need be larger than No. 6, regardless of the size of the wires in your service equipment.

Chapter 9

THREE WIRING SYSTEMS

In residential and farm wiring, three different wiring systems are in common use. They are: (1) nonmetallic sheathed cable, (2) armored cable, (3) metal conduit with four subdivisions: a) rigid conduit, b) intermediate metal conduit, c) thin-wall conduit, and d) flexible conduit. Separate chapters will cover the details of each system. Rigid *Nonmetallic* Conduit is another method, which provides excellent moisture and corrosion resistance, light weight, and ease of installation. See Article 347 in your copy of the Code, as space does not permit a discussion of it in this book. Another system known as Knob-and-Tube is so little used today that it does not warrant space in this book. See Article 324 in the Code.

Which System to Use? Local Codes sometimes prohibit one or more of these systems; what is prohibited in one locality may be required in another. On farms, nonmetallic sheathed cable is used almost exclusively, but it is also used extensively in very many cities, not only in residences but in larger buildings as well. However, in larger cities it is sometimes prohibited for new buildings. Armored cable is being used less and less. Follow local custom; if not sure, consult your power supplier or your electrical inspector.

Many things are done in the same way regardless of which system is used, and these will be explained in this chapter. *Study them well, for they are the foundation for all systems and will not be repeated later.*

All explanations in Chap. 9 to 15 will be for "new work," or the wiring of a building while it is in process of being built. The basic principles for "old work," or the wiring of a building after it is built, will be covered in Chap. 16. Study new work well in order to better understand old work.

Outlet and Switch Boxes: It is not practical to use switches and receptacles fastened to the wall without further protection. Splices in wire and cable, if not properly made, are dangerous. Therefore the Code requires that every switch, every outlet, every joint in wire or cable must be housed in a box. Every fixture must be mounted on a box.* The boxes are called switch boxes or outlet boxes, depending on their particular shape and purpose. Most boxes are of metal with a galvanized finish. However, boxes made of insulating materials are becoming quite common, especially for farm use. They are used only with nonmetallic sheathed cable, and will be discussed in Chap. 17.

Switch Boxes: Every switch and receptacle must be installed in a box; the most common is the type shown in Fig. 9-1. Each box has mounting brackets on the ends, which can be adjusted so that the front of the box will be flush with the surface of the wall. There are two "ears" with tapped holes for screws to hold switches and

---

* Fluorescent fixtures if surface-mounted may have a cable or conduit enter the end of the fixture; if suspended below the ceiling, may be cord-and-plug connected to a receptacle in a ceiling box.

receptacles and similar devices installed in the boxes. The Code requires that such boxes must have a depth of at least $^{15}/_{16}$ in. where a switch or receptacle (device) is to be installed; most are at least 1½ in. deep, but the deeper ones (from 2 to 3½ in. deep) are handier and more generally used. A switch box is sometimes used for fixtures of the wall-bracket type. Locate one at each point where a bracket is to be installed.

Each box holds one device. When two devices are to be mounted side by side, two boxes can be changed into one "2-gang" box by simply throwing away one side of each box, and bolting the boxes together, as Fig. 9-2 shows. Use the screws that are part of the boxes.

Knockouts: Wires and cables must be brought into the box, so "knockouts" are provided; these are sections of metal partially punched out so that they can easily be removed to form openings. A knockout is loosened by placing a screwdriver at the proper point and striking it a stiff blow; use pliers to remove.

Fig. 9-1. A typical switch box, used to house switches, receptacles, etc. The sides are removable.

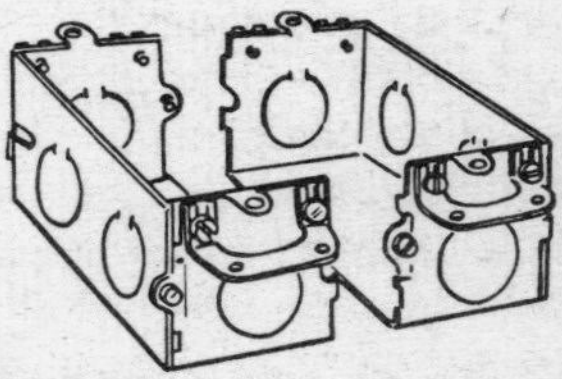

Fig. 9-2. Two single boxes are easily changed into one larger "two-gang" box. Still larger boxes of 3 or 4 or more gangs are made the same way.

Outlet Boxes: Fig. 9-3 shows the most common outlet box. It is octagonal in shape and available in three sizes: 3¼-, 3½-, and 4-in. The 4-in. permits more wires to be used, avoids cramping, and in general reduces the time required for installation. If space permits, use a box at least 1½ in. deep.

Fig. 9-3. The common octagonal box.

Fig. 9-4. Every outlet box must be covered. Above are shown the more common covers.

Outlet boxes must always be covered. A great many different covers are available and common ones are shown in Fig. 9-4. At *A* is shown a blank cover used when a box is merely used to hold a splice or tap. At *B* is shown a drop cord cover; the opening for the drop cord is fitted with a smooth bushing to eliminate sharp edges. At *C* is shown a spider cover used to mount surface-type switches. At *D* is shown a cover with a duplex receptacle. At *E* is shown a keyless receptacle for a lamp and at *F* a similar one with a pull-chain. Many other styles are available.

Other Boxes: If an ordinary switch box is mounted on the *surface* of a wall, as in basement, garage, etc., the sharp corners of the box and the cover are quite a nuisance. In such locations use a box with rounded corners, known as a "handy box" or "utility box." See Fig. 9-5, which shows the box and suitable covers.

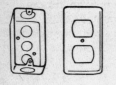

Fig. 9-5. This box is known as a "utility box" or "handy box." Use it for permanently exposed surface wiring as in basements, garages, etc. Neither the box nor the covers have sharp corners.

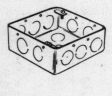
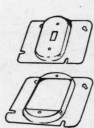

Fig. 9-6. With square boxes, use covers at top when wiring is permanently exposed, those below when wiring is concealed.

Boxes 4-in. square, shown in Fig. 9-6, are used for the same purpose, but are large enough to hold two devices, which require special covers shown in the same picture. They are also used for general-purpose wiring and are especially handy when many wires must enter the same box.

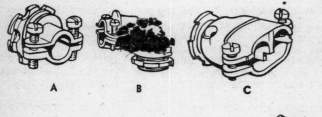

A   B   C

Fig. 9-7. Connectors used in anchoring cable to boxes.

Fig. 9-8. The connector is first attached to the cable by means of a clamp screw.

Fig. 9-9. The connector is anchored to box by means of locknut inside the box.

Number of Wires in Box: The Code in Sec. 370-6(a) specifies the maximum number of wires permitted in a box. Do not crowd a box to its limit, for crowding makes it difficult to do good work, increases the time of doing the work, and tends toward possible shorts and grounds. The Code limits* are as follows:

Kind of Box	Size in Inches	Maximum Number of Wires			
		No. 14	No. 12	No. 10	No. 8
Outlet Box:	4 × 1¼ **Round**	6	5	5	4
	4 × 1½ **or**	7	6	6	5
	4 × 2⅛ **Octagonal**	10	9	8	7
	4 × 1¼ **Square**	9	8	7	6
	4 × 1½ **Square**	10	9	8	7
	4 × 2⅛ **Square**	15	13	12	10
	4¹¹⁄₁₆ × 1½ **Square**	14	13	11	9
	4¹¹⁄₁₆ × 2⅛ **Square**	21	18	16	14
Switch Box:	3 × 2 × 1½	3	3	3	2
	3 × 2 × 2	5	4	4	3
	3 × 2 × 2¼	5	4	4	3
	3 × 2 × 2½	6	5	5	4
	3 × 2 × 2¾	7	6	5	4
	3 × 2 × 3½	9	8	7	6
Handy Box:	4 × 2⅛ × 1½	5	4	4	3
	4 × 2⅛ × 1⅞	6	5	5	4
	4 × 2⅛ × 2⅛	7	6	5	4

* Excerpted with permission from the National Electrical Code, 1981 edition, copyright 1980, National Fire Protection Association, Boston, MA 02210.

The table must be correctly interpreted depending on many factors, as follows:
1. The wires from a fixture to wires in the box are not counted.
2. A wire entering a box and running out again without splice (as is often the case in conduit wiring) is counted as only one wire.
3. Deduct one from the numbers in the table if a box contains a fixture stud, a hickey, or one or more cable clamps (as in Fig. 9-10). The total deduction is one for each type of these items present in the box. Connectors for cable shown in Fig. 9-7 are *not* cable clamps for this purpose.
4. Deduct one from numbers in table for each switch, receptacle or similar device, or a combination of them if mounted on a single strap.
5. A wire originating in a box and ending in the same box (for example the wire from the green terminal of a receptacle, grounded to the box) is not counted.
6. If one *or more* bare grounding wires of nonmetallic sheathed cable enter the box, deduct one from the numbers in the table.

Boxes not listed in NEC Table 370-6(a) [from which the above table is extracted] and nonmetallic boxes will be marked with their cubic inch capacity, and the number of conductors must be calculated on the basis of 2 cu. in. for No. 14; 2¼ cu. in. for No. 12; 2½ cu. in. for No. 10; 3 cu. in. for No. 8; and 5 cu. in. for No. 6, all per Code Table 370-6(b). Use the same factors covered by the notes, above.

Connectors: When cable of any style is used for wiring, the Code requires that it be securely anchored to each box that it enters. There are many kinds of connectors for this purpose; an assortment of them is shown in Fig. 9-7. The connector at *A* is used for ordinary purposes, that at *B* for a sharp 90° turn and that at *C* when two pieces of cable must enter the same knockout. Remove the locknut from the connector. Install the connector on the cable, slip the connector through the knockout, install the locknut inside the box. Be sure to drive the locknut down solidly, so that the lugs on it actually bite into the metal of the box, to form a good continuous ground. See Figs. 9-8 and 9-9. Some boxes have built-in clamps that securely hold the cable entering the box, so that separate connectors are unnecessary. Typical boxes of this kind are shown in Fig. 9-10.

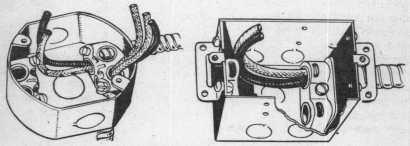

Fig. 9-10. When using boxes of the type above, separate connectors are not used. The boxes have clamps to hold the cable.

Fixture Studs: Small, lightweight fixtures are often mounted directly on outlet boxes, anchored by screws entering the "ears" of the box, as will be explained in Chap. 15. Heavier fixtures need additional support. Install a fixture stud as shown in Fig. 9-11, in the back of the box, using stove bolts. This is unnecessary if a hanger, described later, is used to support the box.

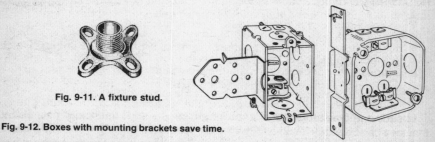

Fig. 9-11. A fixture stud.

Fig. 9-12. Boxes with mounting brackets save time.

Mounting Switch Boxes: Boxes must be firmly secured to studs. Some boxes come with nails attached. If mounting nails pass *through* the box they must not be more than ¼ in. from the back or ends. Another and usually faster way is to use boxes with mounting brackets as shown in Fig. 9-12. Always install boxes so that their front edges will be flush with the outside of the finished wall.

Sometimes a box for a switch or receptacle must be installed *between* studs in a wall, rather than next to a stud. This is difficult to do using a switch box. Use an octagonal or preferably a square box with the proper cover, and a hanger, as will be explained later in this chapter.

Installing Outlet Boxes: If the box is to be installed near a stud, use an outlet box with a mounting bracket; one type is shown in Fig. 9-12. If the box must be mounted between studs or joists, use a hanger of the type shown in Fig. 9-13. It is adjustable in length and has a fixture stud sliding on it. Remove the center knockout in the box, slide the stud to the position wanted, push the stud through the knockout, and drive the locknut on the inside of the box securely home. Then nail the hanger to the studs or joists, at a depth so that the front edge of the box (or the cover mounted on it if of the type shown in Fig. 9-6) will be flush with the wall or ceiling, as shown in Fig. 9-14. Even handier are preassembled boxes with hangers shown in Fig. 9-15.

Recessed Fixtures: If the bottom of your fixture will be flush with the ceiling, it is called a "recessed fixture." See Chap. 15 before installing.

Wiring at Boxes: Regardless of the wiring method used, let from 6 to 10 inches of wire project at each box, the length depending on your choice of two methods of connecting the wire to the terminal screw, as discussed in Chap. 4. Also remember that if a switch or receptacle must be later replaced, wires that may seem just right in the original installation, may be too short for the replacement device which may have terminal screws in a different location.

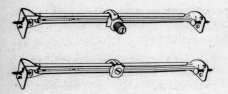

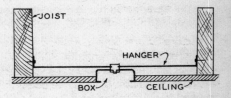

Fig. 9-13. Hangers are adjustable in length.

Fig. 9-14. Hanger with box supported in ceiling.

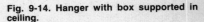

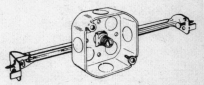

Fig. 9-15. Boxes preassembled with hangers are often used.

Location of Switch Boxes: Boxes for switches should be located near doors, so that they can be easily found as the door is opened. Consider which way the door is to swing; install box on side of door opposite the hinges. Install so that the center of the box is from 44 to 48 in. above the floor.

For receptacles the usual height is from 12 to 18 in. above the floor. Be sure at least one in each room is installed where it is not likely to be hidden behind furniture, so that it will always be easily accessible for vacuum cleaner.

Selection of Switches: While all common switches have the general appearance shown in Fig. 2-1, there are three varieties that you must understand:

Single-pole: used to turn a light on or off from one point. It has *two* terminals and the words ON, OFF, on the handle.

Three-way: used to turn a light on or off from two separate places. It has *three* terminals, and plain handle, without ON, OFF.

Four-way: when a light must be turned on or off from more than two places, use three-way switches at two of the points, and four-way at the remaining points. The four-way has four terminals, and plain handle without ON, OFF.

Types of Switches: Switches are designed for use at an amperage and voltage not higher than the limits stamped into the metal mounting yoke of the switches. Some switches are rated "10A 125V-5A 250V," meaning that the switch may be used to control loads not over 10 amp. if the voltage is not over 125 volts, but only 5 amp. if the voltage is higher but not over 250 volts. Most are rated 15A-120V, or 15A-120/277V. Other switches of course are available for higher amperages.

The most commonly used switch today is called the "ac general-use" type, in the Code. It is very quiet in operation, and long lived. Identify it by the letters "AC" following the ampere and voltage rating on the yoke. As the name implies it may not be used on dc circuits. On ac, it may be used for any purpose except (a) it may not be used to control tungsten-filament lamps (ordinary lamps) at voltages above 120 volts, and (b) if used to control a motor, it may be used only up to 80% of its ampere rating.

Older style switches are called "ac-dc general-use" in the Code. On dc circuits only this style may be used, but they may of course be used on ac circuits also. (If the switch does not have the letters "AC" at the end of its rating, it is an ac-dc type.) If your wiring was installed many years ago, the switches are probably of the ac-dc type, but if one of them fails, there is no need to replace it with one of the same type. Replace it with one of the more readily available ac-only type.

Lighted Switches: Switches are available with a tiny neon light in their handles, the light being *on* while the switch is *off*, to make it easy to locate the switch in the dark. The light has an exceedingly long life, possibly 20 years, and consumes so little power that it costs maybe one cent per year to operate.

Quiet Switches: Ordinary switches make a sharp noise as they are turned on and off. This noise may annoy sick people, wake the baby, and certainly is annoying. The "ac general-use" switch is very quiet. Also available is a completely silent switch, using a bit of mercury in a glass tube, to make and break contact. Mercury switches will not operate unless installed in a vertical position.

Fig. 9-16. Plug-in strip provides outlets every few inches.

Plug-in Strip: Few homes have all the receptacle outlets that the occupant would like. The plug-in strip shown in Fig. 9-16 makes outlets available at intervals of from six to 24 in. as desired. The material consists of a metal channel with wires and receptacles already installed. In living rooms the strip is installed on top of the baseboard, with molding directly above the channel, so that the whole assembly appears to be part of the baseboard. In a kitchen it is installed at a convenient height above the counter, for appliances.

Two-circuit Receptacles: The Code in Sec. 210-70 requires that every livable room, also hallways, stairways, attached garages, and all entrances be equipped with at least some lighting *controlled by a switch;* this includes outdoor lights at doors. This provision is in the interest of safety, to eliminate the danger of accidents caused by stumbling over something in the dark, while trying to find a floor or table lamp. This lighting must consist of permanently installed lighting fixtures in bathrooms and kitchens, but can be from floor or table lamps plugged into receptacles controlled by a wall switch in other rooms.

Duplex receptacles are available that are the equivalent of two separate single receptacles, each with its own terminals. One half of the receptacle is permanently connected for use with clocks, vacuum cleaners, and similar equipment. The other half is controlled by a wall switch so that any lamps plugged into them can be turned off at the same time by the wall switch. Such receptacles are called two-circuit receptacles.

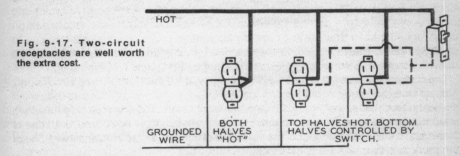

Fig. 9-17. Two-circuit receptacles are well worth the extra cost.

HOT

GROUNDED WIRE — BOTH HALVES "HOT" — TOP HALVES HOT. BOTTOM HALVES CONTROLLED BY SWITCH.

Many brands of ordinary duplex receptacles are so constructed that you can, at the time of installation, change them from ordinary to the two-circuit variety, by breaking out a small brass portion between the two "hot" (brass colored) terminal screws. The wiring diagram is shown in Fig. 9-17.

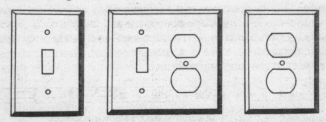

Fig. 9-18. Every switch and receptacle must be covered with a face plate.

Face Plates: Receptacles and switches when installed are covered with face plates of the type shown in Fig. 9-18. These are available in molded plastic in either brown or ivory color, brass in various finishes, glass, and other materials. Use the kind you like best.

Chapter 10

NONMETALLIC SHEATHED CABLE

Nonmetallic sheathed cable costs less than other kinds, is light in weight and very simple to install; no special tools are needed. Its use is limited by the Code to buildings not exceeding three floors above grade. There are two kinds, which the Code calls Type NM and Type NMC.

Type NM Cable, as shown in Fig. 4-5, and described in Chap. 4, may be used only in *normally dry indoor locations.* Type NMC may be used in damp locations and out of doors protected from the weather, and will be described later in this chapter.

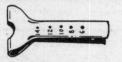

Fig. 10-1. A cable ripper saves time.

Type NM Cable: A jackknife may be used to remove the outer jacket. Cut a slit parallel to the wires, being careful not to damage the insulation of the individual wires. The cable ripper shown in Fig. 10-1 is very handy, and faster than a jackknife. The outer cover is removed for about 8 to 12 inches.

Bending cable sharply may damage the outer cover. The Code says that all bends must be gradual so that, if continued in the form of a complete circle, the circle would be at least 10 times the diameter of the cable. Cable must always run in continuous lengths from box to box — no splices except inside boxes.

Cable must be fastened to the surface over which it runs every 4½ ft., also within 12 in. of every box. Straps of the general type shown in Fig. 10-2 are used for the purpose. While staples are not prohibited by Code, *don't use them* because it is so easy to damage the cable by driving the staples in too hard. A similar caution should be observed regarding connectors and clamps. They should be snug, but not over-tightened to the point where the conductors will be damaged. Sometimes this damage does not show up until some later time, and is difficult to repair once the cable is concealed in the walls.

If when the job is finished the cable will be concealed, it may follow the side of wood members, supported as above outlined, or run through holes bored in the approximate center of studs, joists, or rafters. Where the edge of a bored hole in a stud is less than 1¼ in. from the face of the stud, a ¹⁄₁₆ in. steel plate, or bushing, must be installed to protect the cable against future penetration by nails, etc.

If the finished wiring is exposed (as, for example, in unfinished basements), see to it that the cable is protected against later mechanical injury. The easiest way is to run it along the side of a stud or joist. If run at angles to such timbers (unless the cable runs through bored holes) a running-board must first be installed as shown in Fig. 10-3. The

Code is not specific as to the size of this running-board but the so-called "1 by 2" is fine for the purpose. The cable may never be run across free space, must follow the surface of the building except when mounted on running-boards. In unfinished basements, it may be run through bored holes through the center of joists (cables No. 8-3 or 6-2 or heavier may be mounted directly across the bottoms of joists without a running-board). In accessible attics, cable may run at an angle to joists if protected by guard strips at least as high as the cable, as shown in Fig. 10-4.

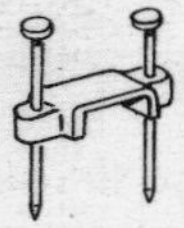

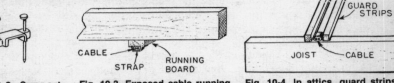

Fig. 10-2. Support cable at least once every 4½ ft. Use straps shown.

Fig. 10-3. Exposed cable running across timbers must be protected by a running board, which prevents damage.

Fig. 10-4. In attics, guard strips may be used. If run through bored holes, the cable needs no further protection.

Plan the Installation: Before you can do any actual wiring, you must make a plan. Decide upon the location of each outlet, each switch. Be generous in the number of outlets and switches; plan an *adequate* installation as described in Chap. 2; remember that it costs very much more to add an outlet later, than it does to include it in the original job.

Cable With Grounding Wire: For most installations using nonmetallic sheathed cable, you will have to use the kind with a bare, uninsulated grounding wire in addition to the insulated wires. Forget this for the moment: pretend you are using ordinary cable without the grounding wire: it will be easier to understand the wiring. *All diagrams in this chapter show connections for the insulated wires only.* Later the connections for the bare grounding wire will be discussed.

Which Outlets on Which Circuit? Your first idea may be to put all the outlets of your basement on one circuit, all the outlets of the first floor on the next circuit, those of the second floor on still another circuit, etc. If you do that and a breaker trips, or a fuse blows, what happens? An entire floor will be dark. Put different parts of any floor on two different circuits; then at least part of each floor will still be lighted even if the rest of that floor is dark.

After you have decided which outlets are to go on each circuit, draw a diagram of each circuit; show how the cable is to run from one outlet to the next, from outlet to switches, and so on. A typical diagram of this kind is shown in Fig. 10-5 which shows 8 outlets, plus 3 single-pole switches and a pair of 3-way switches. The outlets have been labeled *A, B, C, D, E, F, G,* and *H.* Switches have all been labeled *S,* that controlling outlet *B* being indicated as *S-B,* that controlling outlet *G* as *S-G,* that for outlet *H* as *S-H,* and the two 3-way switches controlling outlet *F* as *S-F-1 and S-F-2.*

This diagram still does not tell you how to connect up the wires inside the cable, so that all the parts will work properly. Draw a second diagram and forget that *cable* is being used; show the wires separately just as if they were open wires. Fig. 10-6 shows the same outlets as Fig. 10-5, and each outlet has been labeled the same as in Fig.

10-5. Note that the same scheme has been used as in Chap. 6; a light line like this —— for a white wire, a heavy line like this —— for a black wire (or other color, but not white or green), and a heavy broken line like this - - - for the wire between a switch and the outlet it controls.

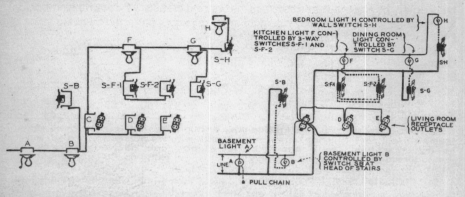

Fig. 10-5. Before proceeding to do any wiring, make a rough layout showing where each outlet is to be located, and how cable is to run from box to box.

Fig. 10-6. The same outlets shown in Fig. 10-5, but now showing how the wires inside the cable are connected. In due course of time, you will be able to get along without the plans.

In the diagram of Fig. 10-6 you will see that outlet *A* (if you disregard the cable that runs on to *B*) is the same as the outlet shown in Fig. 6-3 in Chap. 6, and no detailed explanation is necessary. However, Fig. 10-7 shows the appearance of this outlet completely installed with an outlet box. You have followed the principles you learned in Chap. 6. You have run the white wire from SOURCE to the outlet; you have run the black wire from SOURCE to the outlet.

To better understand the wiring of each additional outlet, consider the cable that runs to it *from the previous outlet,* as the SOURCE for the new outlet. For example, the cable that runs from outlet *A* to outlet *B* becomes the SOURCE for outlet *B*.

In Fig. 10-8, outlet *B* (if you disregard the cable running to *C*) is the same as that in Fig. 6-5, so a detailed explanation may seem entirely unnecessary. However, according to everything you have learned up to this point, and as shown in Fig. 10-6, *both* wires from outlet *B* to switch *S-B* should be *black*, and the 2-wire cable that you are going to use contains one black and one white wire. How then can you comply with the Code? The Code in Sec. 200-7, Exc. 2, makes an exception to the general rule; when wiring with cable (whether nonmetallic or armored) it permits a white wire to be used where a black wire *should* be used — but only in a switch loop, that is, the cable running from an outlet to a switch. It is easy to make the right connections if you remember that each fixture *must* have one white and one black wire connected to it, and if you observe the following simple steps:

1. At the switch, connect the two wires of the cable to the switch.
2. At the outlet, connect the white wire from SOURCE to the fixture, as usual.

3. At the outlet, connect the *black* wire from SOURCE to the *white* wire of the cable that runs to switch; this is contrary to the general rule, but permitted by the Code's exception. This is the only case where a white wire may be connected to a black.

4. Connect the black wire of the cable that runs to the switch, to the fixture as usual.

5. The two wires running on to the next outlet are connected to the two incoming wires (from SOURCE) in the outlet box — black to black, and white to white.

When outlet *B* is properly installed according to these simple rules, it will be connected as shown in Fig. 10-8, which complies with Code Sec. 200-7.

Outlets *C, D,* and *E* are very simply wired as shown in Fig. 10-9. Receptacles have double terminal screws so that the two wires from two different pieces of cable can easily be attached as shown. In the case of *C,* it is necessary to connect three different wires to each side of the receptacle, but there are only two terminal screws. *The Code in Sec. 110-14(a) prohibits more than one wire under one terminal screw.* Splice all the blacks together and all the whites together adding short pieces of insulated wire (called "pigtails"); run those short wires to the receptacle terminal screws. See Fig. 4-15.

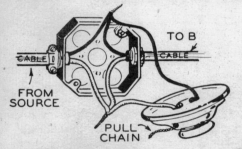

Fig. 10-7. Outlet A of Figs. 10-5 and 10-6, completely installed.

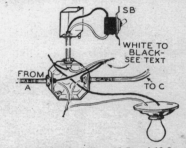

Fig. 10-8. Outlet B of Figs. 10-5 and 10-6: an important diagram. It shows how to connect white wire in cable to switch.

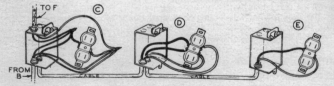

Fig. 10-9. Receptacle outlets are easy to connect, as this diagram shows.

Outlet *F* in Fig. 10-6 (if you disregard the cable running on to the next outlet *G*) is the same as Fig. 6-14 in Chap. 6. Wire it as shown in Fig. 10-10. Run 2-wire cable from the outlet to the first 3-way switch *S-F-1* and 3-wire cable from there to the second 3-way switch *S-F-2*. Again you meet the problem of the proper colors of wire. Remembering the steps outlined in connection with outlet *B,* simply connect the white wire from SOURCE (which in this case is the white wire from outlet *C*) to the fixture as usual. The other wire on the fixture *must* be black, so connect the black wire of the cable that runs on to the first switch *S-F-1*. At the outlet, the white wire in the cable that runs on to the switch, is connected to the incoming black wire from SOURCE, contrary to general rule but permitted by Code exception. The cable that runs on to the next outlet *G* must also be connected; black wire to black and white to white, of the cable from SOURCE (from *C*). This completes the wiring of the outlet *F*; the switches are still to be connected.

Two different cables end in the box for *S-F-1*: one 2-wire cable, one 3-wire cable, five wires altogether. Two of them are white; splice them together so that there will then be a continuous white wire from *F* to *S-F-1* to *S-F-2*, where you connect it to the common or marked terminal of that switch. Connect the black wire in the cable between *F* and *S-F-1* to the common or marked terminal of the first switch *S-F-1*. That leaves two unconnected wires in the cable from *S-F-1* and *S-F-2*: the red and the black. Connect them to the remaining terminals of each switch; it does not matter which color goes to which terminal on the switch. That finishes the wiring.

If you are installing 4-way switches, follow the diagrams of Figs. 6-18 and 6-19. White wire may, per the Code exception, be used where otherwise a wire of a different color would be required.

Outlet *G* is exactly the same as outlet *B*; wire it in the same way.

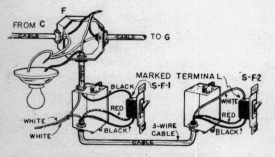

Fig. 10-10. The wiring of outlet F of Figs. 10-5 and 10-6 is shown here, with the two 3-way switches that control the outlet. This is the diagram for any outlet controlled by 3-way switches, when using cable. Study carefully the colors of the wires in the cable. Compare it with Fig. 6-14.

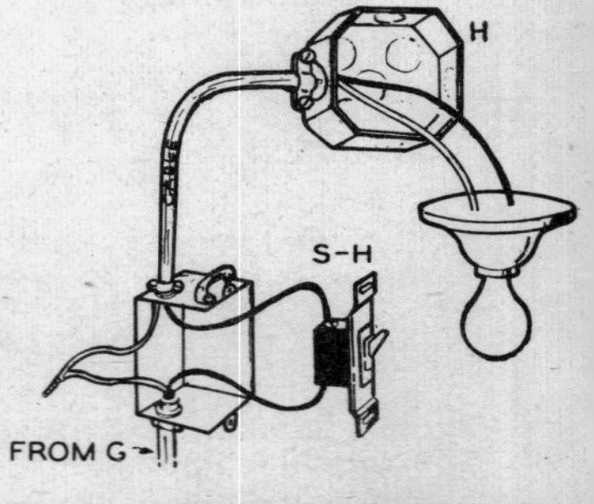

Fig. 10-11. Sometimes the cable from SOURCE does not run first to the outlet, then to the switch. Use this diagram when cable runs first to the switch, then to the outlet.

Feed Through Switch Box: In all the outlets wired so far, you have run the cable first to the outlet box and fixture, then on to the switch. When you come to outlet *H* of Fig. 10-6, you will see that the cable runs first to the switch box *S-H* and then on to outlet *H*. This combination is even simpler to wire than the others, for there is no problem with the colors of the wire, as you can see from Fig. 10-11, which pictures this outlet completely wired.

Outlets Beyond H: You can *not* install additional outlets beyond *H* by simply connecting the black and white wires of the cable for the new outlet, to the black and white wires in outlet *H*, because then the new outlet would be turned on and off by switch *S-H*. However, you can add an additional outlet by tapping in at the switch *S-H*; splice the wires of the cable for the new outlet, to the incoming cable from *G*, as shown in Fig. 10-12. Another way is to run 3-wire cable from *S-H* to *H*; splice the two wires in the cable for the new outlet, to black and white in *H*, as shown in Fig. 10-13.

A cable to any additional outlet can be spliced to any existing cable in any outlet box by splicing white to white, black to black, *if each wire can be traced all the way back to* SOURCE *without interruption by a switch.*

Substituting 3-Way Switches for Single Pole: It is a very simple matter to substitute two 3-way switches for a single-pole switch, in any wiring diagram. Study Fig. 10-14 — the starting point is an outlet already wired with two wires ready for a switch. If a single-pole switch is to be used, connect it to the two ends of the wires as at *A*. If 3-way switches are to be used, substitute the combination of *B*.

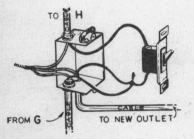

Fig. 10-12. One method of adding an outlet beyond an existing outlet.

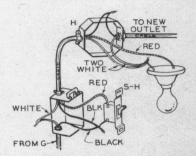

Fig. 10-13. Another method of adding an outlet beyond an existing outlet.

If you want to add a switch to a diagram which shows an outlet permanently connected, without a switch, cut the black wire. That gives you two ends of black wire to which the switch connects, or two ends to which you will splice the 2-wire cable which runs to the switch.

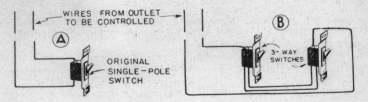

Fig. 10-14. It is a simple matter to substitute a pair of 3-way switches for a single-pole switch in any wiring diagram. Simply substitute the right-hand combination for that at left.

Junction Boxes: Sometimes it is necessary to make a T connection in cable, when there is a long run and no convenient outlet from which to start the T branch. Use an outlet box, run the three (or more) ends of cable into it, splice all black wires, all white; cover with a blank cover, and the job is finished. See Fig. 10-15. Such junction boxes must always be located *where permanently accessible.*

Type NMC Cable: The ordinary Type NM cable is entirely suitable in normally dry locations, but when exposed to moisture or corrosive vapors it rots away. For that reason a different kind of cable which the Code calls Type NMC was developed. As you will see in Fig. 10-16, the individual insulated wires are imbedded in a solid sheath of plastic material; sometimes there is a glass overwrap on each insulated wire. This construction does away with all fibrous materials such as paper, jute, and fabric sheaths of ordinary Type NM, so that there can be no rotting. Therefore the Type NMC may be used in moist, damp or corrosive locations indoors or out, but not exposed to the weather, and not buried in the ground.

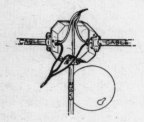

Fig. 10-15. A junction box contains only the splices of several lengths of cable.

Accordingly, you *may* use it instead of Type NM anywhere, but *must* use it in damp or corrosive locations such as barns and other farm buildings, damp basements, and so on. Install it as you would ordinary Type NM.

In some localities it may be difficult to locate Type NMC cable. In that case use Type UF, which is very similar, doesn't cost a great deal more, and may be used wherever Type NMC may be used. In addition, it may be used in wet locations or buried directly in the ground. Type UF is discussed in more detail in Chap. 17.

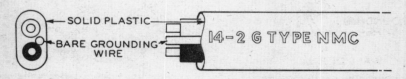

Fig. 10-16. Nonmetallic sheathed cable, Type NMC, may be used in dry or damp locations.

Use Cable With Grounding Wire: Types NM and NMC are made both with and without a bare, uninsulated grounding wire in addition to the two or three circuit wires, but are seldom seen without the grounding wire. See Fig. 4-5 and 10-16.

To avoid unnecessary complications as the subject was discussed, everything in this chapter up to this point has been based on using cable *without* the grounding wire. But cable *with* the grounding wire must be used in most cases. The transition from one kind to the other is not at all difficult.

At one time cable with the grounding wire was required only to boxes containing receptacles. It must now run to practically every box. Since all receptacles in a dwelling must be of the 3-wire grounding type, and all exposed metal parts of lighting fixtures must be grounded, the only locations where cable without a grounding conductor could be used would be to a non-metallic fixture (such as a keyless porcelain socket) or to a switch in a nonmetallic box, with a non-metallic cover plate held on by nylon screws. So, the practical thing is to use only cable with a separate bare (or green) grounding conductor to *every* box. The difference in cost does not justify keeping track of two kinds of cable on the job, and the risk of using the wrong one.

At the starting point in the service equipment, connect the bare grounding wire to the neutral strap, or separate grounding bus, in the cabinet. If the circuit originates in a separate panelboard, *not part of the service equipment,* the bare grounding wire *must* be connected to a separate equipment grounding bus, which is grounded to the cabinet, and *not* to the neutral, for the neutral is insulated from the cabinet at such locations. At all boxes connect the insulated wires as you would if no grounding wire were involved. At each outlet or switch box, the bare wire must be properly connected. Assume that two ends of cable enter the box; you will then have two ends of bare wire. Cut another piece of bare wire a few inches long, or use a commercially available green insulated, or bare, pigtail assembled to a screw. Connect all three ends solidly together; a solderless connector of the type that was shown in Fig. 4-18 will serve the purpose. Ground the opposite end of the short wire to the box itself. You may use one of the clips shown in Fig. 10-17, or install an extra screw in one of the unused holes in the box. That screw may be used *only* for grounding the bare wire.

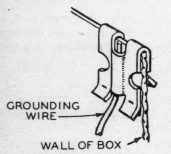

GROUNDING
WIRE—

WALL OF BOX

Fig. 10-17. The clips shown are very convenient in grounding to boxes.

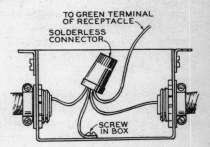

TO GREEN TERMINAL
OF RECEPTACLE
SOLDERLESS
CONNECTOR

SCREW
IN BOX

Fig. 10-18. How to install the bare grounding wire of nonmetallic sheathed cable.

If a receptacle is to be installed in that box, run a bare wire (or insulated *green* wire) from the *green* terminal of the receptacle, to the junction of the other bare wires; your solderless connector will then join four ends of bare wire. See Fig. 10-18. However, many brands of receptacles are now available, suitable for use without the grounding wire from the green terminal, if installed in *metal** boxes. If a receptacle is installed in a surface-mounted box such as a handy-box in a basement, so that the mounting strap or yoke is in good solid contact with the box, the wire from the green terminal is not

* When using nonmetallic boxes, there is no need to connect the bare wire to the box, but all the bare wires must be connected to each other, and to the green terminal of the receptacle.

required. However, regardless of circumstances, the bare wires in the cables must always be connected to each other, and to the box. Be sure to fold all the bare wires well back into the box so they cannot later touch a live terminal on the receptacle.

In any event, regardless of the kind of boxes used, the grounding wire must be so installed that it will be continuous from box to box all the way back to the service equipment. Removing a receptacle or other device from the box must not interrupt the continuity of the wire. All this is in the interest of good grounding, which means a greater degree of safety.

Testing: Even experienced electricians with confidence in their work test their installations. You should test your work before it is covered up so that any corrections can be made while it is still accessible. Proceed as outlined in the following paragraphs, *before* the power supplier has connected to your installation.

All you need is a couple of dry cells and a doorbell. Be sure all of the permanent splices are completed. Also, at every switch point, have the wires touching just as if the switch were there and in the ON position.

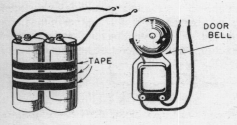

Fig. 10-19. To test the finished wiring, all you need is a doorbell and a couple of dry cells.

When you have done all this, connect two dry cells in series as shown in Fig. 10-19, leaving a hook on the end of each wire. Hook these wires to the white and black wires where the circuit originates (service, or panelboard). You will then have a 3-volt current on that circuit, where you will later have 120-volt current. Touch the doorbell across each pair of wires at the various outlet boxes where a fixture or receptacle is later to be connected; if the wiring is correct the bell will ring. If the outlet is controlled by a switch, the bell should stop ringing when the two switch wires are separated. Repeat for each circuit. If everything checks, and assuming you have used *nonmetallic* boxes, the wiring has been properly done and can be finished as outlined in Chap. 15.

If you used *metal* boxes, one more test is necessary at each box where you have a bare grounding wire connected to the box. Connect the doorbell between the black wire and the box itself. If the bare grounding wire has been properly installed, the bell will ring; the white ground*ed* wire and the ground*ing* wire are connected to each other at the service equipment.

Chapter 11

ARMORED CABLE

Armored cable is quite simple in construction, as Fig. 11-1 shows. The wires are Type T or TW, each wrapped in a spiral layer of tough paper. The galvanized steel armor is strong but quite flexible. Between the paper and the armor there is a bare aluminum bonding strip. This together with the armor itself serves the same purpose as the bare grounding wire in nonmetallic sheathed cable. The armor itself is a grounding conductor, but the armor is made of steel which is a very poor conductor compared with copper; the turns do not make good electrical contact with each other. Hence the need for the bonding strip to assure low resistance and a good ground.

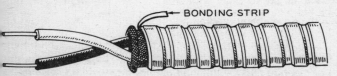

Fig. 11-1. Armored cable consists of two or three Type T or TW wires, protected by a layer of tough paper and flexible galvanized steel armor. Note the bonding strip under the armor.

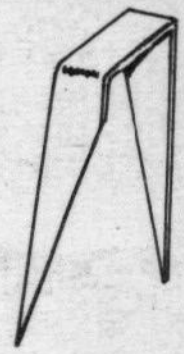

Fig. 11-2. Staple for cable.

Armored cable may be used only in permanently dry locations; don't use it outdoors. Like nonmetallic sheathed cable, it must be supported every 4½ ft. and also within 12 in. of each box. It may be supported using straps similar to those used with nonmetallic sheathed cable; staples shown in Fig. 11-2 are more frequently used. The Code requires rust-proof staples. Drive them home with a hammer, but not hard enough to damage the cable.

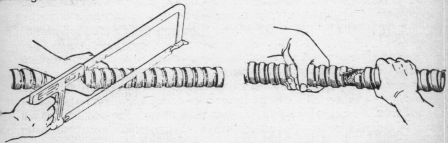

Fig. 11-3. Note the proper angle of the blade in sawing armored cable. Be careful not to damage the wires.

Fig. 11-4. A twist will remove the cut end of the armor. Let about 8 in. of wire stick out beyond the armor.

Avoid sharp bends; the Code says that the bends must be such that if completed into the form of a complete circle, the circle would be at least 10 times the diameter of the cable.

Cutting Cable: Use a hacksaw; hold it in the position shown in Fig. 11-3, almost at a right angle to the strip of armor, rather than the cable itself. Be most careful to saw only through the armor, without touching the insulation of the wires, or the bonding strip. This is not too easy and if you have not had experience, make some practice cuts on odd pieces of cable before proceeding to do wiring. Bending the cable sharply at the point where it is to be cut is helpful in sawing. Actually it is difficult to saw completely through only a single turn of armor, but usually if the center part of the strip is sawed completely through and the edges partly through, a sharp bend will break the armor. A sharp twist on the two pieces of armor as shown in Fig. 11-4 will remove the short end. The cut should be made about 8 in. from the end of the piece, so that you will have plenty of wire in the box for connections.

Fig. 11-5. Always use a fiber anti-short bushing shown at right, at the end of the cable, between wires and the armor. It provides protection against the sharp edges of the armor, which might puncture a wire and cause a ground, or a short-circuit.

Bushings: Cut a length of cable, examine the cut end of the armor; you will find sharp jagged edges on the armor, pointing inward toward the wire. These teeth tend to puncture the insulation of the wire and cause short circuits and grounds. The Code requires that a bushing of tough fiber must be inserted at the end of the cable, between the steel armor and the wires. Such bushings are supplied with the cable and Fig. 11-5 shows one of them.

Fig. 11-6. In removing paper, first unwind a few turns inside the armor.

Fig. 11-7. A sharp yank tears off the paper inside armor, makes room for the bushing.

Fig. 11-8. Insert the bushing between wires and armor.

Fig. 11-9. A properly-installed bushing.

At first you may find it hard to insert the bushing because there is so little room under the armor; the paper wrapped around the wires is in the way. So, you must make room for the bushing. First unwind the paper a few turns *under* the armor, then give it a sharp yank and it will tear off *inside* the armor, all as shown in Figs. 11-6 and 11-7. Then insert the bushing, as shown in Fig. 11-8. Figure 11-9 shows a cross-section of a piece of cable with the bushing properly inserted.

Connectors: When you cut cable, be sure you do not cut off the bonding strip under the armor. Let it project an inch or more beyond the armor.

Fold this grounding strip back over the outside of the cable, then insert the fiber insulating bushing as described in the preceding paragraph, and then install a connector, as was shown in Chap. 9, in the steps of Figs. 9-8 and 9-9 (unless you are using boxes with cable clamps, which make the use of cable connectors unnecessary). The connectors used with *armored* cable are similar to those used with nonmetallic sheathed cable except that at the end which goes into the box, there are "peep-holes" through which the red color of the fiber bushing can be seen; this allows the inspector to see that the bushings have been installed. If you do not use the bushings, your job will not be passed by the inspector.

Using the Cable: After you have prepared the ends of the cable with bushings and connectors, install the cable in the same way as described for non-metallic sheathed cable in the preceding chapter. The problems concerning correct colors of wire are exactly the same as with nonmetallic sheathed cable. As with nonmetallic cable, splices are not permitted; the cable must be in one piece from box to box. If a splice is necessary, make it in a junction box as shown in Fig. 10-15.

Be specially careful in anchoring the connectors to outlet and switch boxes. The connector must be tightly clamped to the armor of the cable; be sure the aluminum bonding strip of the cable is bent back over the outside of the armor, and solidly clamped, by the connector. In other words, don't let the strip lie in a part of the connector that doesn't solidly clamp it against the armor. After inserting the connector into the knockout of the box, drive the locknut down tightly enough to bite down into the metal of the box. This makes a good electrical connection so that current can flow through the armor from box to box. This happens only in case of accidental grounds. The white wire in the cable is grounded; the grounding strip is also grounded. If the black wire at some point where the insulation is removed accidentally touches the armor or the box, it is the same as touching the white wire. That is a ground fault and causes the fuse protecting that circuit to blow, a signal that something is wrong.

Receptacles: There is no need for cable with an extra grounding wire as with nonmetallic sheathed cable; the armor of the cable and the grounding strip under the armor serve the same purpose. But the green grounding terminal of the receptacle must still be *effectively* grounded.

If the mounting yoke of the receptacle is in firm solid contact with the box, as in surface wiring, no further action is necessary. In flush work the plaster ears on the yoke of the receptacle usually prevent it from resting directly on the box so that the only metallic contact between the box and the receptacle is through the small mounting screws, and that is not good enough. Unless you use the specially approved receptacles mentioned in Chap. 10, you must install a short length of wire from the green terminal of the receptacle, to the box, using either of the methods discussed in the preceding chapter concerning nonmetallic sheathed cable.

Testing: Test as with nonmetallic sheathed cable using metal boxes.

Chapter 12

CONDUIT

The four most commonly used types of conduit will be discussed here: 1. The original type is known as rigid metal conduit, is generally made of steel but may be made of aluminum or other metals, and has approximately the same dimensions as standard weight water pipe, size for size. 2. Intermediate metal conduit (IMC) has slightly smaller wall thickness and larger inside diameter than rigid metal conduit, and is made only of steel. 3. Thin-walled conduit (EMT), called "electrical metallic tubing" by the Code, has still smaller wall thickness; in sizes ½-in. through 1½-in. it has the same inside diameter and in sizes 2-in. and larger it has the same outside diameter as rigid metal conduit. 4. Flexible metal conduit, commonly called "Greenfield" or "flex." Space does not permit a full discussion of other types of conduit such as Liquidtight Flexible Metal Conduit, for which see Article 351 in the Code, or PVC (polyvinyl chloride) and other types of Rigid Nonmetallic Conduit, for which see Article 347.

Rigid Conduit: Standard sizes are ½, ¾, 1, 1¼, 1½ and 2-in., plus larger sizes used mostly in commercial work. The actual *inside* diameter is a little larger than the trade sizes indicated above. Conduit differs from water pipe in several important ways. It is carefully inspected to make sure that it is entirely smooth inside to prevent damage to the wires as they are pulled into the pipe; it has a rust-resistant finish both inside and outside. Each length bears an Underwriters' label. It may be used indoors or outdoors, but may require supplementary protection in corrosive locations such as in cinder fill.

Fig. 12-1. Rigid conduit looks like water pipe but differs in several important aspects. It comes only in 10-ft. lengths and each length bears the Underwriters' label.

Fig. 12-2. A conduit bender is needed to make good bends in conduit. The one shown here is for thin-wall conduit. Instructions usually come with the bender. →

Bending Conduit: If conduit is bent sharply, it will collapse. Bends must be gentle and gradual, so that the internal diameter will not be reduced at the bend. The Code requires bends which if continued into the form of a complete circle, would be not less than 8 in. in diameter for the half-inch size; 10 in. for the ¾-inch; 12 in. for the 1-inch; 16 in. for the 1¼-inch; 20 in. for the 1½-inch; 24 in. for the 2-inch; 30 in. for the 2½-in. To do a good job, use a bender similar to the one shown in Fig. 12-2. For larger sizes, use factory-bent elbows.

Cutting and Threading Conduit: The preferred way to cut rigid conduit is to use a hacksaw, using a blade with 18 teeth to the inch. Cutting will leave a sharp edge or at least burrs at the cut, which might damage wires as they are pulled into the conduit. Use a pipe reamer to remove these dangerous projections. Thread the conduit using dies similar to those used with water pipe, with a taper of ¾ in. to the foot.

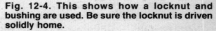
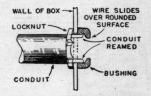

Fig. 12-3. Conduit is fastened to boxes by means of locknuts and bushings, both shown above.

Fig. 12-4. This shows how a locknut and bushing are used. Be sure the locknut is driven solidly home.

Installing Locknut and Bushing: Conduit is anchored to a box by means of a locknut and bushing, both of which are shown in Fig. 12-3. Their use is made clear from Fig. 12-4 which shows how the locknut is used on the outside of the box, the bushing on the inside. The locknut is not flat, but has teeth on one side; the side with the teeth faces the box. The bushing has a rounded surface on the inside diameter, over which the wires slide while being pulled into the pipe. To install properly, first screw the locknut on the pipe as far as it will go, then slide the pipe through the knockout in the box, then install the bushing on the inside of the box. Screw the bushing on tight, as far as it will go, and only then tighten up the locknut on the outside of the box, running it home tight so that the teeth will dig down into the metal of the box, to provide good continuous ground. Many electricians prefer, and some local codes require, that two locknuts be used, one on each side of the box wall, plus a bushing. The Code requires double locknuts with circuits of over 250 volts to ground.

Intermediate Metal Conduit: Cutting, threading and bending are done the same as for rigid metal conduit, except that benders providing side wall support, similar to the EMT bender shown in Fig. 12-2, *must* be used, as the thinner walls must be supported to prevent collapse during bending. Threaded ends will appear a little different, as the tops of the first few threads will be flat, but properly cut threads will mate with standard conduit fittings. In general, IMC is recognized for the same uses as is rigid metal conduit.

Thin-Wall Conduit: This material is properly known as electrical metallic tubing, or EMT. It is shown in Fig. 12-5; each length bears an Underwriters' label. It may be used either indoors or outdoors. In trade sizes ½ in. through 1½ in. it has the same *inside* diameter as rigid conduit. The outside diameter is much less than that of rigid conduit, because the wall is much thinner; as a matter of fact, the wall is so thin that it cannot be threaded. Lengths are coupled together and connected to boxes with special pressure fittings — coupling and connector of one type are shown in Fig. 12-6. Be sure the end of the tubing goes all the way into the connector, against the shoulder stop. Tightening the nut securely clamps the tubing into the fitting. In sizes 2 in. and larger EMT has the same *outside* diameter as rigid conduit.

Cutting and Bending Thin-Wall Conduit: A hacksaw with 32 teeth to the inch is the most convenient tool for cutting thin-wall conduit. It must be reamed after cutting. Bend it like rigid conduit.

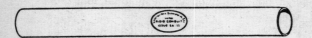

Fig. 12-5. Thin-wall conduit cannot be threaded; it is much lighter than rigid conduit. It comes only in 10-ft. lengths.

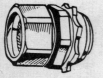

Fig. 12-6. Thin-wall conduit is anchored to boxes using the connector shown at left. The picture also shows a coupling.

Size Conduit to Use: Whether thin-wall or rigid conduit is used, the Code in Tables 1, 3A, 3B and 3C of Code Chap. 9 restricts the number of wires that may be used inside each size of conduit. The size pipe required for each combination is shown here:

Size of Wire	Number of Wires to be Installed				
	2	3	4	5	6
14	½	½	½*	½*	½*
12	½	½	½*	½*†	½ ▲ †
10	½	½*	½*	½ ▲ †	¾*
8	½*†	¾*	¾*†	1*	1*†
6	¾*	1*	1*	1¼	1¼*
4	1*	1*	1¼*	1¼*	1½*
2	1*	1¼	1¼*	1½*	2
1/0	1¼*	1½*	2	2*	2½
2/0	1½*	1½*	2*	2*	2½*
3/0	1½*	2	2*	2½*	2½*

Conduit sizes in table above are for Type T or TW wire (also Type RHW unless the size is marked *, ▲, or †). If you are using Type RHW wire, note that this is available two ways: with and without an outer cover. The kind you are using will affect the size of conduit required:

* If RHW *with* an outer cover, use conduit one size larger than shown.

▲ If RHW *with* an outer cover, use conduit two sizes larger than shown.

† If RHW *without* an outer cover, use conduit one size larger than shown.

Installing Conduit: Whether rigid or thin-wall conduit is used, the procedure is the same. The pipe is first installed, the wire pulled into it later. No run (as the conduit from box to box is called) may have more than the equivalent of four quarter-bends in it. All wires must be continuous, no splices being permitted inside the pipe. Support the conduit using pipe straps. It must be supported within 3 ft. of every box, plus additional supports not over 10 ft. apart for ½- and ¾-in., 12 ft. for 1-in., 14 ft. for 1¼- and 1½-in., and 16 ft. for 2- and 2½-in. For larger sizes see Table 346-12 in your copy of the Code.

Until you are well experienced, you may find it best to bend the conduit first, then cut to the required length, for the finished piece after bending may turn out to be a fraction of an inch too long or too short.

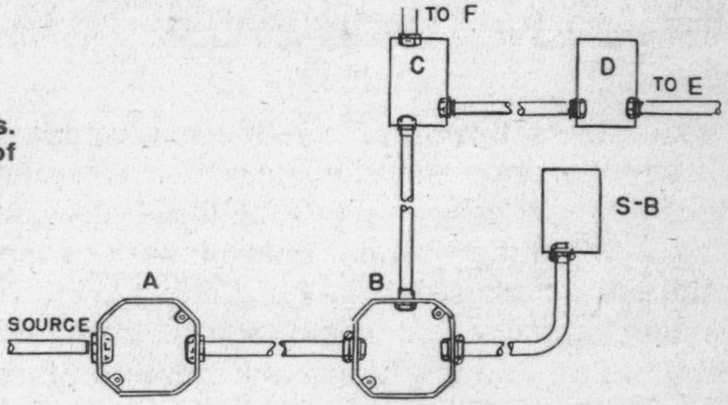

Fig. 12-7. Outlets A, B, C, and D of Figs. 10-5 and 10-6, installed with runs of conduit, ready for wires to be pulled in.

Pulling Wires Into Conduit: Assume that you have installed the conduit for outlets A, B, C, and D of Fig. 10-6 in Chap. 10. The installation will look as shown in Fig. 12-7. You are ready to pull the wires into place. For a short run with just two small wires, they can probably be simply pushed in at one end and through to the next outlet. If the runs are longer or with bends, fish tape must be used: a highly tempered steel tape about ⅛ in. wide and 1/16 in. thick. As purchased, it will have a loop formed at one end. The tape is flexible enough so that it will go around corners, but stiff enough so that it will not

TAPE OPTIONAL

Fig. 12-8. Fish tape makes it easy to pull wires into conduit.

buckle when pushed into conduit. Push the tape into the conduit until the end with the loop appears where the wires are to enter the conduit. Push the bare wires through the loop of the tape, then twist them back upon themselves, as shown in Fig. 12-8. The use of the insulating tape shown is optional; it does help to hold the wires fast. Should the loop on the end of the tape break, you can't form a new one by just bending, because the highly tempered tape would break. Heat the end of the tape with a blowtorch, let it cool; this will soften the end so that a new loop can be formed.

In pulling the wires into outlets of Fig. 12-7, pull in short pieces from A to B, more separate pieces from B to S-B, from B to C, and from C to D. But if the wires from A to B to C for example, did not require connections at B, you would feed a length in at A, through B, and on to C. In any event let from 8 to 12 in. stick out at each box where a connection is to be made. Outlet S-B will contain a switch; use only black wire, for there are no exceptions as to color, as in cable wiring.

Receptacles: Ground the green terminal to the box, unless you are using the newer type of receptacles mentioned in Chap. 10, that makes this unnecessary.

Testing: Test as when using armored cable.

Fig. 12-9. Flexible conduit is installed in same way as armored cable, but wires are pulled into place later.

Flexible Metal Conduit: This material is commonly called "Greenfield" and is similar to the armor of armored cable, but without wires, and as shown in Fig. 12-9. Use only indoors in permanently dry locations (unless using wire approved for wet locations: one having a "W" in the type designation, as for example, Type TW). The smallest for general use is trade-size ½-in. but is about ¾ in. in outside diameter. Greenfield is not commonly used for a complete wiring system, except in a few communities, but is widely used where some flexibility or movement is required, such as at motors. Install it as you would armored cable, using connectors of proper size, then pull the wires into place. You must also install a ground*ing* wire, which may be bare, or insulated if green, except that in lengths up to 6 ft., where the conductors are protected at 20 amp. or less, the flex may serve as the equipment grounding means without an additional grounding wire. The size of the grounding wire depends on the rating of the fuse or breaker protecting the hot wires. If 15-amp., use No. 14; if 20-amp., use No. 12; if 30- to 60-amp., use No. 10; if 100-amp., use No. 8; if 200-amp., use No. 6. This wire is connected just as the bare grounding wire in nonmetallic-sheathed cable. Ground the green terminals of receptacles as with nonmetallic-sheathed cable. Test as when using nonmetallic-sheathed cable with ground wire.

Depth of Boxes: In installing any kind of conduit, be sure to use boxes deep enough so that the knockouts on the sides or ends of the box are close enough to the back, so that there will be room for the conduit behind the thickness of the wall or ceiling finish material.

Chapter 13

MISCELLANEOUS WIRING

Previous chapters have outlined the most common wiring problems, using the devices ordinarily used. In this chapter will be discussed some wiring problems that do not readily fit into another chapter, and some devices not previously mentioned. Study this chapter well for it will enable you to include in your installation some of the niceties that are needed for a really modern wiring job.

Interchangeable Devices: In Fig. 13-1 is shown a face plate with three openings, and an assortment of devices such as switches, receptacles and pilot lights. The face plate is supplied with a metal strap with three openings, on which you can mount any three devices in any combination you want. The three devices can then be installed in an ordinary single-gang box. Using three *ordinary* devices, you would need a 3-gang box which costs more, requires more space and much more installation time. Often the use of these devices permits locating three devices in a location where there isn't room for a 3-gang box, and makes a neater installation in any case. Caution: Remember that the number of wires entering any box is limited; review the table in Chap. 9. When installing three devices on a single strap, the number of wires entering an ordinary box may exceed the permitted maximum. In that case, use a 4-in. square box and a cover as shown in Fig. 9-6.

Plates are also available with only one or two openings, so that uniformity of appearance may be preserved throughout an installation.

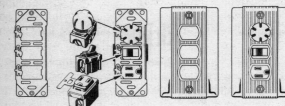

Fig. 13-1. Interchangeable devices permit three switches, receptacles and the like, to be installed in a single-gang box.

Fig. 13-2. Mount your clock on top of a clock-hanger receptacle.

Clock Receptacles: Everybody uses electric clocks, but in many rooms and especially in the kitchen, long unsightly cords to the clock are untidy and dangerous. Get rid of those cords by installing a special receptacle shown in Fig. 13-2, where the clock is to be located. The receptacle fits down into the box on which it is mounted; locate the clock on top of the receptacle. Shorten the cord on the clock so it will be only a few inches long, and it and the receptacle are completely concealed behind the clock.

Dimmers: Some people like to be able to control the brightness of the lights in parts of their homes, such as dining rooms or recreation rooms. This can now easily be done by using special dimming switches, most of which can be used only with incandescent lamps, not fluorescent. Some types however can be used with fluorescent lighting. In each case, remove the ordinary switch, in its place install the dimming switch. Two kinds are available.

One is a quite inexpensive switch that provides HIGH-OFF-LOW positions, and controls up to 300 watts. It can be used to replace only an ordinary single-pole switch, not a 3-way. The somewhat more expensive type controls the brightness continuously from OFF to full bright. Some models are available in the 3-way type, replacing one of a pair of 3-way switches.

Doorbells and Chimes: These necessary devices are very easy to install. The power is furnished by a transformer, which is a device which reduces 120-volt ac to a different voltage, still ac. For doorbells this is usually about 6 to 10 volts. Doorbell transformers have two "primary" *wire* leads which are permanently connected to any point on any 120-volt circuit that is *not* controlled by a switch, and two "secondary" *screw terminals* for the low voltage. They are so made that the power consumed while the bell is not ringing is under one watt despite the permanent connection to the line. When you push the button to ring the bell, you close the low-voltage circuit, and then the transformer draws about five watts from the 120-volt line. Transformers operate only on ac.

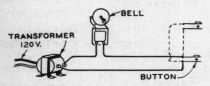

Fig. 13-3. The wiring of a doorbell with a transformer is very simple, as this diagram shows.

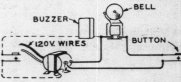

Fig. 13-4. Adding a buzzer for the back door to the diagram of Fig. 13-3 is easy, as shown above.

The installation is very simple. Consider the two secondary or low-voltage terminals on the transformer as the SOURCE for a circuit, the bell as the outlet, and the button as a switch. Breaking it down this way makes all these diagrams very simple. Figure 13-3 shows the installation of one bell with a push button in one location. If a button is wanted in an additional location, install it as shown in the dotted line, in parallel with the first button.

In most installations, a single bell is not considered enough. The usual system is to install a bell for the front door, a buzzer for the back door. See Fig. 13-4, and you will see that it is the same as Fig. 13-3 for a single bell, with the buzzer added as shown in the dotted lines.

Instead of doorbells, musical chimes are more often used. One of them is shown in Fig. 13-5. They are so designed that when the front door button is pushed, two musical notes are sounded, and when the back door button is pushed, only a single note is sounded. No matter how long the button is held down, the sound does not repeat. The wiring of these devices is the same as for the doorbell.

If you install chimes in place of an existing doorbell, you may find that the sound is not very loud. You will have to install a new transformer, for those used with doorbells usually deliver only about 6 to 10 volts, and many chimes require more nearly 20 volts, or thereabouts.

Because the usual transformer used for doorbells and chimes cannot deliver more than about 20 volts, nor more than about five watts, there is little danger of shock or of fire even in case of accidental short-circuit. Therefore the wire used for this low-voltage wiring does not need much insulation, and bell wire of the kind shown in Fig. 13-6 is commonly used; it is insulated only with a thin layer of plastic. Attach it to the surface over which it runs using insulated staples.

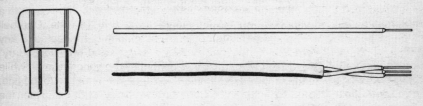

Fig. 13-5. Chimes are rapidly replacing bells.

Fig. 13-6. Shown above are wires for connecting doorbells and chimes. Because of the low voltage involved, such wires need little insulation.

Low Voltage Switching: In this method of wiring, the 120-volt wires run to various outlets where power is consumed, in the usual fashion. They *do not* run to the individual switches.

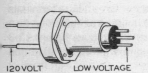

120 VOLT   LOW VOLTAGE

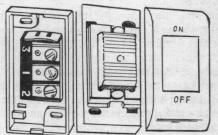

ON

OFF

Fig. 13-7. The low-voltage (small) end of this relay projects out of the outlet box.

Fig. 13-8. The switches used in remote control wiring are shown in two views above, with cover at right. The same kind of switch is used at every location; use as many switches as you wish to control one outlet (or several outlets if you wish).

In each outlet box where power is consumed a very small relay (which is an electrically operated switch) is installed with one end projecting out of the outlet box through a knockout. A 24-volt transformer is installed in some convenient location, and furnishes power to operate all the relays. Because the voltage is low and because the transformer delivers only a very few watts even if short-circuited, the wiring between the transformer, the relays and the individual switches, resembles doorbell wiring. Inexpensive wire, insulated for only a low voltage, is run from the transformer to each relay, then to as many switches as you wish. The wire does not need to run through conduit nor does it have to be cable of the armored or nonmetallic sheathed type. Run it or fish it through walls, staple as required. The switches do not need outlet boxes. Each switch is really a double pushbutton. Pushing the top end of the button on any switch momentarily, turns on the light; pushing the bottom end momentarily on any switch, turns it off.

Figure 13-7 shows one of the relays, Fig. 13-8 one of the switches, and Fig. 13-9 the general type of wire used. Fig. 13-10 shows the wiring circuit.

Fig. 13-9. There are only 24 volts on the control wires in remote control wiring, so inexpensive wire is used.

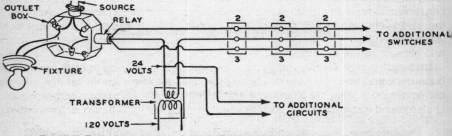

Fig. 13-10. Typical remote control circuit. One transformer is used for the entire house. The same circuit is often used for remote control yard lights.

In installing the relays you have a choice of two methods. In the first, run the low-voltage wires to each box where a relay is to be installed, and let at least 6 in. project into the box, through the knockout in which the relay is to be later installed. In the second method, install the relays in the boxes first, and then connect the low-voltage wires to them later; be sure to allow at least 6 in. slack in those wires. Unless the extra length of wire is provided, you would later find it impossible to replace a relay should one prove defective. As an alternate to installing relays at the controlled outlets, relays can be ganged in a centrally located box which has separate compartments for the circuit wiring and the low-voltage wiring.

Garage Wiring: If the garage is attached to your house, treat it as another room. It will be found to be a great convenience to install 3-way switches, one at the garage door, the other at the door between garage and house. Provide a couple of GFCI protected receptacles for battery charger, tools, etc. [One is *required* by Code Sec 210-52(f).] If the garage is separate from the house, proceed as outlined in the following paragraphs.

The minimum wiring in a garage is a light, turned on or off by only a switch in the garage, and fed by a pair of wires from the house. Most people will demand a light that can be turned on and off from either the garage or the house, which requires a 3-way switch at each end, wired as shown in the basic diagram of Fig. 13-11. Three wires are required between the house and the garage.

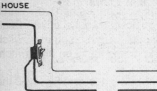
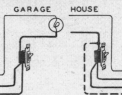
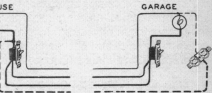

Fig. 13-11. Diagram for garage light fed from house, with 3-way switch at each end.

Fig. 13-12. If a receptacle that is to be permanently ON is to be added, four wires are needed.

Every garage should have a receptacle outlet. Although the Code requires it only in an *attached* garage, you would be well advised to install GFCI protection for any receptacles in a *detached* garage, to provide an additional margin of safety. It is very convenient to have this receptacle always live, instead of being turned on and off with the garage light (so that you can plug in a battery charger and have it work all night without having the garage light on all night.) That requires four* wires between house and garage. Follow the diagram of Fig. 13-12.

Underground or Overhead Wires? Underground wiring used to be the exception because it was very expensive. Today inexpensive cable is available that may be buried directly in the ground, and as a result underground wiring is now quite common. Underground installations will be discussed in Chap. 17, which covers farm wiring.

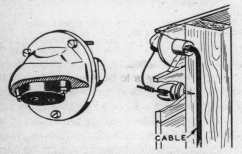

Fig. 13-13. The easiest way of bringing wires into a garage or similar building, is to use the special fitting shown.

Fig. 13-14. Another method of entering a building.

If you are going to use overhead wiring, there are several ways in which the wires can be brought out of one building and into another, for example house and separate garage. Fig. 13-13 shows a very convenient entrance cap requiring only a single hole through the wall; use it at both ends. It is suitable for use with cable as pictured; you can also use conduit by running it directly into the fitting, which accommodates ½-in. conduit. Fig. 13-14 shows another way of using the ordinary entrance cap (as used for service entrances), a short piece of conduit and an outlet box inside the building.

* There is a "trick" circuit using only three wires but it does not meet Code requirements.

Three-wire Circuits: An ordinary 2-wire 120-volt circuit consists of two wires: the grounded wire and a hot wire as shown in Fig. 13-15. When two such 2-wire circuits are run *to the same general area,* there is a total of four wires as shown in Fig. 13-16 in which wires *B* and *C* are the grounded wires, the wires *A* and *D* are hot wires, *connected to opposite sides* of SOURCE, so that the voltage between *A* and *D* is 240 volts, but between *A* and *B,* also between *C* and *D,* is 120 volts. But note that the two grounded wires *B* and *C* are connected together at the neutral busbar in your service equipment.

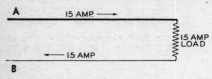

Fig. 13-15. One 2-wire 120-V. circuit carrying 15 amp.

Since they are connected together at the starting point, they become in effect one wire. Why use two? There is no need for two; use one wire and instead of two 120-volt 2-wire circuits you will have one 3-wire 120/240-volt circuit as shown in Fig. 13-17. Be sure that each receptacle is connected to the neutral wire to obtain 120 volts; if connected to the two hot wires the voltage of course would be 240 volts.

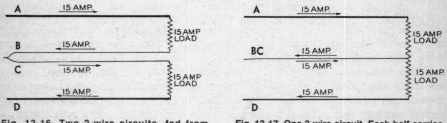

Fig. 13-16. Two 2-wire circuits, fed from opposite legs of 120/240-V. service, each carrying 15 amp.

Fig. 13-17. One 3-wire circuit. Each half carries 15 amp. but the neutral BC carries no current.

Installing a 3-wire circuit not only saves material, but also reduces the voltage drop, for if the loads on each half of the circuit (between *A* and *BC,* or between *D* and *BC*) are equal, the wire *BC* carries no current at all. Important: In Fig. 13-16, each wire *B* and *C* is merely a *grounded* wire, but in Fig. 13-17 the single wire *BC* will, if both loads are equal, carry no current at all, so it becomes a grounded *neutral* wire. If the loads are unequal, it will carry only the difference between the current in *A* and that in *D,* but it is still a neutral. One 3-wire circuit is especially recommended for the small appliance circuits discussed in Chap. 5, in place of two 2-wire circuits. But, watch two points most carefully.

Wires *A* and *D must* be connected to opposite legs of the incoming service-entrance wires. If you connect *A* and *D* to the same leg, the wire *BC* instead of carrying a small number of amperes (or no current at all if the loads between *A* and *BC,* and *D* and *BC,* happen to be identical) would be carrying a double load, possibly as much as 40 amp. if No. 12 wire is used; its insulation would be damaged and might start a fire. Remember the grounded wire is not protected by a fuse or breaker.

The neutral must be specially watched. A receptacle has two terminal screws on each side. When wiring 2-wire circuits, it is the usual practice to end two different grounded wires at these two terminals, as is done especially with cable wiring. In wiring *3-wire* circuits, this must *not* be done. If using conduit, run a continuous wire with a loop as shown in Fig. 4-17. When using cable, there is no way of making a loop and you must follow the procedure that was shown in Fig. 4-15. Connect the ends of the white wires of two lengths of cable to each other and to another short piece of white wire, the opposite end of which you then connect to the whitish terminal screw of the receptacle. Unless this precaution is observed, removing a receptacle for replacement would (while there is no connection to the receptacle) result in a break in the neutral wire. In turn this would place all the receptacles connected to one leg (beyond the receptacle temporarily removed) in series with those connected to the other leg, *all at 240 volts.* Appliances would malfunction and be ruined or at least badly damaged. Lamps would burn out.

The split-wired receptacle, shown in Fig. 9-17, can also be installed on a three-wire circuit, wired with one half on phase A, the other on phase D (Fig. 13-17), and a common neutral. This is often done in the kitchen, where *two* small appliance receptacle circuits are required. Be sure that the branch circuit disconnecting means at the panelboard opens both hot wires with a single operation of the hand, so that future maintenance at these outlets can be done safely.

Smoke Detectors: Self-contained smoke detectors which will sound an alarm when visible or invisible products of combustion are sensed have been credited with saving many lives in dwellings by waking the occupants early enough for them to escape from a fire. Many building codes require a smoke detector in the hallway outside bedrooms, or above the stairway leading to bedrooms on an upper floor. Battery-operated, plug-in, and direct-wired models are available. The required location will be on the ceiling, or on the side wall not more than 12 in. from the ceiling. Be sure not to place the detector in the path of ventilation which will move air past the detector faster than in other parts of the room. If direct-wired, choose a circuit with often-used lights on it, such as the bathroom light, to assure the circuit could not be off without it being noticed, but wire it directly across the two circuit wires, unswitched.

Track Lighting: Decorative fixtures to "wash" walls with light, highlight art objects, or illuminate wall-hung paintings, can be installed anywhere along a surface-mounted (or flush) track. Many types of fixtures are available for a great variety of decorative lighting functions. Power to the track is supplied at one end through a ceiling outlet.

Outdoor Receptacles: Every house is required by Code Sec. 210-52(d) to have at least one outdoor receptacle. It is better to install two, one on the front and one on the back, for appliances, yard tools, Christmas tree lights, etc. For installation methods, see Chap. 17. Remember that each outdoor receptacle (or the circuit on which it is installed) must be protected by a GFCI as discussed in Chap. 7.

Heavy-Duty Receptacles: The ordinary receptacle is rated at 15 amp. Receptacles with 20-, 30-, and 50-amp. ratings are described in the next chapter.

Chapter 14

WIRING OF HEAVY APPLIANCES

Appliances may be classified into three groups: portable, fixed and stationary. It is sometimes difficult to decide to which group a given appliance belongs; common sense must rule. Portable appliances are those that are or can be readily moved about in normal use. Fixed are those that are permanently fastened or secured in one place; oil burner motors and water heaters are examples. Stationary are those that are not secured or fastened in one place, but are not moved about during normal operation; electric ranges and clothes dryers are examples for they normally remain where installed, but when moving from one house to another they can be easily moved with other household goods. The Code uses these terms to classify appliances: "permanently connected"; "cord-and plug-connected"; "fastened in place"; and, "located to be on a specific circuit".

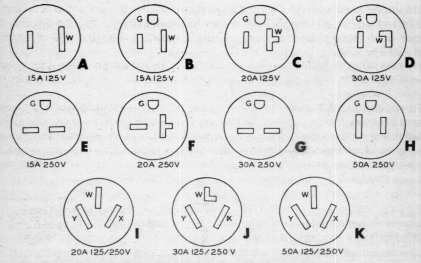

Fig. 14-1. An assortment of receptacles. Those rated at 15- and 20-amp. are 1.327 in. in diameter; those rated at 30- and 50-amp. are 2.12 in. in diameter.

Receptacles: Receptacles are rated in amperes and volts. Those rated at 125 volts may be used at any voltage not over 125. Those rated at 250 may be used only at voltages over 125, but not over 250. Those rated 125/250 volts may be used only for appliances that operate at 120/240 volts and which require a neutral wire running to the appliance.

Figure 14-1 shows a variety of receptacles, labeled A, B, C and so on to K. Except A, all 125- and 250-volt receptacles have a third opening for the third prong on a 3-prong plug, which is connected to the equipment grounding wire, as discussed in Ch. 7. The 125/250-volt receptacles have a third opening for the third prong on the plug, but that is connected to the grounded neutral wire of the circuit.

Note that a plug that will fit A will also fit B or C; a plug made for B will also fit C; but a plug made specially for C will not fit A or B. All of the other configurations are non-interchangeable.

At A is shown the former "garden variety" of 15-amp. receptacle (usually made in the duplex configuration) and now used only for replacements. The receptacle shown at B is the same as at A, except it has the third opening for the third prong on a 3-prong plug, for the equipment grounding wire. At C is shown a similar 20-amp. receptacle. At D is shown a 30-amp. receptacle for larger 120-volt loads.

At E, F, G, H are shown respectively 15-, 20-, 30- and 50-amp. receptacles for loads operating at 240 volts. Note that each is provided with an opening for the equipment grounding prong on a 3-prong plug. Note that that prong and that opening must never be used for a neutral wire if the appliance operates at 120/240-volts, rather than at 240 volts.

At I, J and K are shown 20-, 30-, and 50-amp. receptacles for appliances operating at 120/240 volts. The 30-amp. is used mostly for clothes dryers, and the 50-amp. for ranges. Such receptacles must never be used for loads operating at 240 volts rather than 120/240 volts.

In all the illustrations, the opening marked "G" is for the equipment grounding wire; the one marked "W" is for the white circuit wire; those marked "x", "y" or "z" are for the hot circuit wires.

Do not imagine that these are the only available receptacles. There are literally dozens of others, in 2-wire, 3-wire, 4-wire and even 5-wire types. Besides the ordinary variety there are others so designed that the plug cannot be removed without first twisting the plug to unlock it. (Some special-purpose types not used in ordinary wiring are designed so that only a plug and receptacle of the same brand will fit each other.)

Fig. 14-2. Ranges are connected to 50-amp 125/250-volt receptacles of the type shown here. The pigtail is connected to the range, then plugged into the receptacle.

Receptacles come in a variety of mounting methods to fit various boxes and plates, flush and surface types. The 50-amp. receptacle is shown in both the surface-mounting type and flush-mounting type in Fig. 14-2, which also shows a typical plug with "pigtail" cord attached, to fit. These are used mostly in the wiring of electric ranges. The similar 30-amp. receptacle is used for clothes dryers.

Individual Circuits for Appliances: The Code rules as to when an appliance requires an individual branch circuit, serving no other load, are quite complicated. In general, you will be following the Code rules if you provide a separate circuit for each of the following:

1. Range (or separate oven or counter-mounted cooking units).
2. Water heater.
3. Clothes dryer.
4. Automatic clothes washer.
5. Garbage disposer.
6. Dishwasher.
7. Any 120-volt "fixed" (permanently-installed) appliance rated at 12 amp. (1440 watts) or more. This includes motors.
8. Any "fixed" (permanently-installed) 240-volt appliance.
9. Any automatically-started motor such as an oil burner, pump, and so on.

Disconnecting Means and Overcurrent Protection: Every appliance must be provided with some means of disconnecting it completely from the circuit, and also with overcurrent protection.

Portable Appliances: A plug-and-receptacle arrangement is all that is required. Of course the plug and receptacle must have a rating in amperes and volts at least as great as that of the appliance.

Small Fixed Appliances: If the appliance is rated at 300 watts or less (⅛ hp. or less) no special action need be taken. The branch circuit overcurrent protection is sufficient. No special disconnecting means is required.

Larger Fixed and Stationary Appliances (non-Motor): If the appliance is connected to a circuit *also serving other loads* and if the circuit is protected by a circuit breaker or by fuses mounted on a pullout block, no special action need be taken. If the circuit is protected by plug fuses, you must install a separate switch of the general type shown in Fig. 17-6, for each appliance. The switch need not be fused, but unfused switches are hard to find, so it is customary to use the fused kind, with one fuse for 120-volt appliances and two fuses for 240-volt appliances. A circuit breaker may be used instead. Do not install a fuse or breaker in the grounded wire.

Larger Appliances, Motor-Driven: If the motor is not *automatically* started, proceed as if the appliance were *not* motor-driven. But if the motor is automatically started such as a water pump or an air conditioner, then follow the same rules except that built-in or separate motor overload protection rated at not over 125% of the ampere rating of the motor must be provided, to protect the motor against overload or failure to start (See Chap. 18). While not required by Code, it is wise to provide an individual circuit for each automatically-started motor.

Wiring 240-volt Appliances with Cable: White wire may be used only for the grounded wire. But that wire does not run to any appliance operating at 240 volts. Therefore the wires running to a 240-volt load may be any color except white or green. Now when you use a 2-wire cable to connect a 240-volt load, the cable contains one black wire, and one white wire, but the white wire may not be used. What to do? Follow Code Sec. 200-7, Exception 1. Paint each end of the white wire black, or wrap black tape around the ends, and the cable will be considered as having two black wires.

Bear in mind that an electric range does not operate at strictly 240 volts. When any burner is turned to "high heat" it operates at 240 volts; when turned to "low heat" it operates at 120 volts. In other words it is a combination 120/240-volt appliance, therefore all three wires including the neutral must run to it.

Wiring Methods for Heavy Appliances: The Code does not restrict the methods to be used; use conduit or cable as you choose. However, the Code makes one important exception. While many appliances must be grounded, in the case of ranges (including counter units and separate ovens) and dryers, their frames may be grounded to the neutral circuit conductor, provided it is No. 10 or heavier. See Code Sec. 250-60. Moreover, for these appliances and no others,* you may use service entrance cable with a bare neutral provided it runs from the appliance directly to the service equipment.

If the appliance is to be connected by cord and plug, run your cable up to the receptacle, which may be either flush-mounted or surface-mounted. The Code demands that the receptacle be located within 6 ft. of the intended location of the appliance; with good planning it should be possible to locate it even closer to the appliance, thus simplifying the installation of the appliance.

Wiring of Ranges: As already mentioned, any particular burner of a range operates at either 120 or 240 volts, depending on whether it is turned to LOW, MEDIUM, or HIGH heat. The individual burners are so connected within the range that it is impossible for the neutral wire to carry as many amperes as the two hot wires. For that reason, the wires to the range usually include a neutral one size smaller than the hot wires. For most ranges, two No. 6 plus a No. 8 neutral are used; for smaller ranges, two No. 8 with a No. 10 neutral are occasionally used.

Run your circuit up to the range receptacle of Fig. 14-2; this is rated at 50 amp., 125/250-volts. The range is connected to the receptacle using a pigtail cord shown in the same illustration. This also serves as the disconnecting means.

The Code requires that the frame of the range be grounded, but does not require a separate grounding wire. The range is grounded through the neutral wire, when you use one of the pigtail cords and range receptacles just described, and the bonding strap is connected between the neutral wire and the frame. If local codes do not permit grounding to the neutral, the bonding strap from the frame to the neutral must be removed, and a 4-wire cord and plug used.

Sectional Ranges: The trend is away from complete self-contained ranges consisting of oven plus burners, towards individual units. The oven is a separate unit, installed in or on the wall where wanted. Groups of burners in a single section are installed in or on the kitchen counter where convenient. All this of course makes for a very flexible arrangement, and permits you to use much imagination in laying out a modern, custom-designed kitchen. The Code calls such separate ovens "wall-mounted ovens," and the burners "counter-mounted cooking units." Here they will be referred to merely as ovens and cooking units or counter units.

* Ranges and dryers are 120/240-volt appliances, and the neutral wire carries current in normal operation. Three-wire service-entrance cable with a bare neutral may also be used in wiring 240-volt appliances such as water heaters, etc. to which the neutral does not run, provided the bare wire of the cable is used *only* as a grounding wire.

Self-contained ranges are considered *stationary* appliances by the Code; ovens and cooking units are considered *fixed* appliances. They may all be either permanently connected, or cord-and-plug connected.

Two basic methods may be used in the wiring of ovens and cooking units. The more economical method is probably to supply a separate circuit for the oven, another for the cooking unit. The alternate method is to install one 50-amp. circuit for the two combined. Any type of wiring method may be used and by special Code exception [Sec. 338-3(b)] service-entrance cable with a bare neutral may also be used. Regardless of the wiring method used, the frame of the oven or cooking unit must be grounded. For ranges, ovens and cooking units only, the Code permits the frame to be grounded to the neutral wire serving the unit, provided only that the wire is No. 10 or larger, and runs directly to the service equipment if uninsulated.

Assuming that a separate circuit is installed for the oven, proceed as outlined earlier in this chapter for fixed appliances. Use wire with the ampacity required by the load. The oven will probably be rated about 4500 watts, which at 240 volts is equivalent to about 19 amp., so No. 12 wire would appear suitable, but the minimum is No. 10 because of the grounding requirement. At the oven, the circuit wires may run directly to the oven, but some prefer to install a pigtail cord and a receptacle. Do note that the plug will *not* serve as the disconnecting means, as it does when installing a self-contained range because the plug and receptacle are concealed behind an appliance fastened in place.

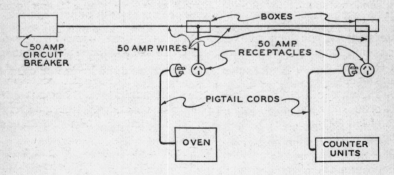

Fig. 14-3. It is best to provide separate circuits for oven and counter units, but both may be placed on one circuit as shown.

So far as the cooking units are concerned, proceed exactly as for the oven, again using a minimum of No. 10 wire; this has ampacity of 30 amp. and will provide a maximum of 7200 watts, which will take care of most cooking units. Use a pigtail cord and a receptacle if you wish for ease in servicing the unit.

If you install a single circuit for oven and cooking units combined, it must be a 3-wire 50-amp. circuit. Any wiring method may be used, including service-entrance cable with bare neutral. The receptacles may be only the 50-amp. type, and may be flush receptacles installed in outlet boxes, or the surface type shown in Fig. 14-2. The circuit will be as shown in Fig. 14-3, which should be self-explanatory. The wires to the receptacles must be the same size as the circuit wires. However, the wires from the receptacles to the oven or cooking unit may [per Code Sec. 210-19(b) Exception 2] be smaller, provided they are heavy enough for the load, not smaller than No. 12 (No. 10 if used for grounding), and not longer than necessary to service the appliance. Do note that receptacles are not required; they may be found convenient in installation. The oven or cooking unit may be connected directly to the circuit wires, naturally within junction boxes; again the Code exception will permit smaller wires between the junction box and the appliance, under the conditions already discussed.

Clothes Dryers: A dryer is a 120/240-volt appliance. Wire it as you would a range. Use a 30-amp. receptacle as shown at J of Fig. 14-1, and a pigtail cord similar to that used for a range but with smaller wires.

The Code requires that the frame of the dryer be grounded, but Sec. 250-60 permits it to be grounded to the neutral of the three wires to the dryer; it must be No. 10 or larger. Use any wiring method you wish, conduit or cable. The same section permits you to use service-entrance cable with a bare neutral (as in range wiring), provided it runs from the service equipment directly to the dryer receptacle. The plug and receptacle serve as the disconnecting means.

Automatic Washers: The Code considers the automatic washer to be a stationary appliance, which is certainly logical since the connection to plumbing prevents easily moving it from one place to another. Washers are equipped with a cord and plug for easier servicing, so provide a 20-amp. grounding receptacle on a 20-amp. circuit. No separate switch is required.

Water Heaters: The Code, in Sec. 220-2(a), requires that loads which are expected to be on continuously for three hours or more not exceed 80 percent of the branch circuit rating. Dwellings and farms seldom have such continuous loads, but a domestic water heater is required to be so considered. This means that a 4500 watt water heater must be on a 25 amp. circuit, not a 20 amp. (4500 watts ÷ 240 volts = 18.7 amp., but 18.7 = 93.5 % of 20. The next higher standard overcurrent device rating is 25 amp., and 18.7 = 75% of 25.) This additional 25% of the continuous loads must be added to feeder and service calculations also [NEC Sec. 220-10(b), 230-41(a)].

Look at the water heater terminal box for a marked temperature rating for the branch circuit wires. If there is no marking, the circuit wiring can be T or TW (or other 60°C wire) but if marked 75°C then use a wire with an H in its designation (or HH if marked 90°C). For further information, see Code Article 310. Where higher temperature wires are required, it is common to splice the ordinary branch circuit wires in a junction box near the water heater to short lengths of the higher temperature wire extending to the heater.

In some localities power for heating water is sold at a reduced rate, the heater being connected to the circuit through a special electrically operated switch furnished by the power supplier, which connects the heater to the power line only during "off-peak" hours. As a result, water cannot be heated for several periods of several hours each, on any one day. If your installation is of this type, do the wiring as already described, except that the wires should start from the power supplier's time switch, instead of from your service equipment.

Grounding: In Chap. 7 are listed the specific appliances that must always be grounded, per Code Sec. 250-45(c). This is especially important if the appliance is installed where a person can touch the appliance, and also the ground or a grounded object. (A concrete floor even if tiled as in a recreation room is considered the same as the actual earth.) For safety's sake always ground your appliances.

How to ground ranges, ovens, cooking units and dryers has already been discussed. Other heavy appliances must also be grounded. If the appliance is supplied with a cord that includes a grounding wire, and a plug with a grounding blade fitting a properly-installed grounding receptacle, that is all that is required.

If there is no cord and plug but the circuit wires run directly to the appliance, and if the wiring is in conduit or using armored cable, check to make sure the frame of the appliance is grounded to the junction box on the appliance, to which conduit or armor is anchored. But if the wiring is by nonmetallic sheathed cable, you must use cable with the bare grounding wire, which must be connected to the frame of the appliance.

In the case of a water heater on a farm, if the wiring is grounded to a driven ground rod, but there is some buried water pipe no matter how short, you *must* interconnect such pipe with the ground rod. This is an essential step to prevent a difference of voltage between them, and to minimize danger from lightning. This will be discussed in more detail in Chap. 17.

Chapter 15

FINISHING AN INSTALLATION

All the wiring described up to this point is done as the building progresses. Switches, receptacles, face plates and fixtures are usually installed only after the walls are papered or painted. This final portion of the work is only a very small part of the total and is quite simple.

Installing Wiring Devices: Figure 15-1 shows how to mount a switch and its plate on a switch box. The switch is mounted in the switch box using machine screws that come with the switch. The plate in turn is mounted on the switch using screws that come with the plate. Inspect any switch or receptable; you will see that the holes for the screws that hold the device to the box are not round but elongated. They are elongated so that the device can be mounted straight up and down. See Fig. 15-2 which shows a box installed crookedly, but with the receptacle straight up and down.

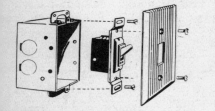

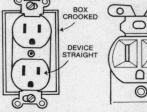

Fig. 15-1. The switch or other device is first mounted in the box, the plate then fastened to the device.

Fig. 15-2. Devices can be mounted straight in a crooked box because of oval holes in strap.

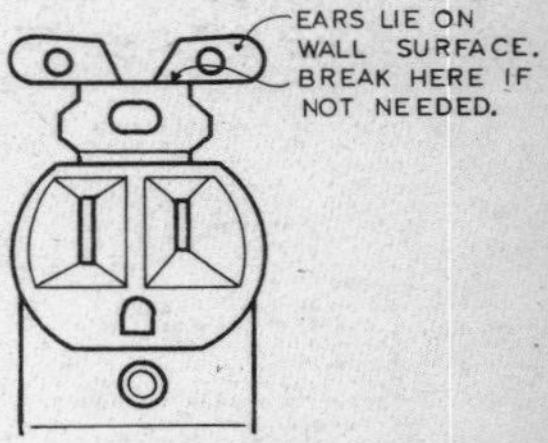

Fig. 15-3. Plaster ears are helpful in aligning a device flush with the wall surface.

Boxes must be installed so that the front edges are no more than ¼ inch below the front side of the finished wall or ceiling, *provided* that these consist of noncombustible materials (materials that will not burn); in all other cases they must be flush with the wall or ceiling surface. If the front edge of the box is not flush with the surface, and you install a switch, receptacle or other device with its mounting yoke touching the box, it will be so deep inside the wall that it will not protrude far enough through the openings in the plate when it is finally put on. This not only makes an untidy job but also makes it hard to operate switches. For this reason switches and similar devices have "plaster ears" as shown in Fig. 15-3, which will lie on top of the wall surface and bring the device to the proper level, even if the front edge of the box is below the level of the surface. You can easily break off the ears if for any reason they are in the way. Sometimes cheaper devices do not have plaster ears, but are provided with spacing washers: these are placed between the device and the box, to bring the device to the proper level, so that the plate will fit properly.

Here is a tip which can save you a lot of trouble: After connecting the wires to a switch or receptacle, fold the wires around to tightly hug the back of the device. Then when pushing the device into place the resistance offered by the stiffness of the wires is taken by the back of the device and not by the wiring terminals. Many loose connections, especially at receptacles, result when an acceptably tightened terminal screw is loosened as the device is pushed into place in the box.

Mounting Face Plates: When mounting face plates do not draw up too tightly on the mounting screws. The most common plates are made of plastic and are quite easily damaged. Plates for duplex receptacles have only a very narrow strip of material between the two openings, and if you pull down too tightly on the screw you will crack this bridge and ruin the plate.

Hanging your Fixtures: On some fixtures one wire is white, the other black. More usually both are of the same color, but one has a colored tracer thread woven into the covering of the wire. The white wire, or the wire with the tracer, always goes to the white wire in the box. The other wire goes either to the black wire in the box, or to the switch. Connect the wires from the fixture to the wires in the box, using solderless connectors.

The Code requires that all fixtures be mounted on outlet boxes. The very simple fixtures can be mounted directly on the boxes using screws supplied with the fixture. This method is shown in Fig. 15-4. Somewhat larger fixtures often use a special strap supplied with the fixture, and this method should be clear from Figs. 15-5 and 15-6. (For fluorescent fixtures, see Chap. 9).

Fig. 15-4. Small fixtures can be mounted directly on the outlet box.

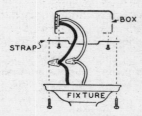

Fig. 15-5. Sometimes a strap is first mounted on the box, the fixture mounted on the strap.

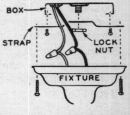

Fig. 15-6. Another method of mounting a fixture.

Still larger fixtures are commonly hung directly on a fixture stud mounted in the back of the outlet box, or on the fixture stud which is part of the hanger on which the box is supported. The "stem" of the fixture is threaded to fit the fixture stud. Slide the canopy down the stem; when the work at the outlet box is finished, all connections made, slide the canopy up to conceal the wiring. See Fig. 15-7.

For mounting a wall bracket, many install a switch box just as if a switch or receptacle were being installed, but this is not recommended. Use a larger box, for example, the type shown in Fig. 9-3 or 9-6. With the wall bracket you will find a mounting strap, a short threaded nipple, and a knob (or a long heavy screw in place of the nipple and knob). Mount the strap on the box, then install the fixture on the strap. Sometimes a fixture stud is mounted in the back of the box, in place of the strap. If the wall is combustible (if it can burn, as in the case of wood) the space from the box to the edges of the fixture, per Code Sec. 410-13, must be covered with noncombustible material such as sheet metal. With some pan-type fixtures (a pan-type fixture mounts on the surface of the ceiling, and extends beyond the outlet box) the noncombustible material, usually fibreglass, comes with the fixture.

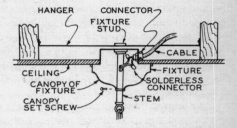

Fig. 15-7. The fixture is supported on the fixture stud. The stem of the fixture fits the stud.

Thermal insulation installed above the ceiling can prevent the dissipation of heat from a fixture, especially a pan-type incandescent lamp fixture. A surface type fixture should not be installed on a thermally insulated ceiling unless the fixture is marked "Type I.C." (Insulated Ceiling) as part of the required UL marking.

Recessed Fixtures: A portion of a recessed fixture must be installed at the "rough-in" stage of the work, rather than after the ceiling finish is on and painted. Most residential recessed fixtures have a connection box spaced away from the fixture housing itself, to allow connection to ordinary branch circuit wiring. However, some will be marked with a temperature rating for the branch circuit wiring which may require that you use a wire with an H or HH in its type designation (TH, THW; 75°C: THHN, XHHW; 90°C). Be sure the fixture is spaced at least ½ in. away from the joists or other combustible material (except at the points of support) and that it will not be blanketed by thermal insulating materials which will prevent the escape of heat (unless marked "Type I.C.," meaning it will not overheat even when blanketed with thermal insulation.) Do not install lamps in such fixtures larger than the size marked on the fixture, as the additional heat will damage wire insulation, and could start a fire.

An outdoor porch fixture if installed on a ceiling is installed just as an indoor fixture. If it is exposed to the rain, install it as in indoor work but be sure the fixture is one designed for exposure to the weather, make sure it has a UL label, and that rain cannot penetrate behind the fixture and enter the building walls.

It is impossible to describe all possible methods of mounting a fixture; it all depends on the construction of the particular fixture. With the help of the general information given here, you should have no trouble, for fittings to suit the particular fixture involved are usually supplied by the manufacturer.

Chapter 16

OLD WORK: MODERNIZING

Everything in previous chapters describes "new work," or the wiring of buildings while they are being built. This chapter will describe the wiring of buildings after they have been completed; this is known as "old work." There is very little difference between the two, except that in old work there are a great many problems of carpentry. The problem is to cut an opening where a fixture is to be installed, another opening where a switch is to be installed, and then to get the cable inside the wall from one opening to the other — with the least amount of work, and without tearing up the walls or ceilings more than is necessary.

One house to be wired may be five years old, another may be a hundred years old; different builders use different methods of carpentry. All this means that every job will be different. No book can possibly describe all the methods used and all the problems that you will meet. Watch buildings while they are being built to get an idea of construction at various points. In old work good common sense is of more value than many pages of instruction.

In general, old work requires more material because it is often wise to use ten extra feet of cable to avoid cutting extra openings in the walls, or to avoid cutting timbers. In locating outlets, bear in mind that all wires must be fished through walls and ceilings. Sometimes by moving an outlet or a switch a foot or so, a difficult job of boring through joists or other timbers can be avoided. Many problems can be solved without cutting any openings except the ones which are to be used for outlet boxes and switch boxes. Others require temporary openings in the wall, which must later be repaired.

Systems Used in Old Work: The rigid conduit or EMT systems cannot be used unless the building is being practically rebuilt. Use nonmetallic sheathed cable with grounding wire, or armored cable, as is the custom in your locality. In a few areas flexible metallic conduct is used; be sure to install the grounding wire along with the circuit wires, as discussed in Chap. 12.

Fig. 16-1. The Code permits boxes as shallow as ½ in., the type shown here.

Code Requirements for Old Work: Boxes enclosing flush devices must be at least ¹⁵⁄₁₆ in. deep. Lighting outlet boxes may be as little as ½ in. deep, as shown in Fig. 16-1, but deeper boxes should be used wherever possible. Cable is simply pulled into the walls and anchored to the outlet and switch boxes. Each piece must be a continuous length from box to box.

Temporary Openings: Sometimes a temporary opening must be made in a wall, so that cable can be fished around a corner. On papered walls, use a sharp safety-razor blade, cut through the two sides and the bottom of a square. Soak the cut portion with a wet rag, which will soften the paste. Lift the cut part off the wall, bend it upwards, using the top edge as a hinge. Use a thumb-tack to hold it in place, out of the way. All this is shown in Fig. 16-2. When the wiring is finished, paste the paper back into place.

Openings for Boxes: In all wiring the wall material must come up close to the box. It is not practical to cut an opening of the exact size of the box, so it will be necessary, after the wiring is finished, to replace the wall material so that it will come up close to the box. Use prepared patching plaster.

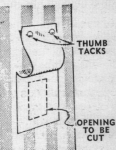

THUMB TACKS

OPENING TO BE CUT

Fig. 16-2. In old work, lift the wallpaper over the area of a temporary opening, as shown here. The top of the paper serves as a hinge.

In old work the switch and outlet boxes are supported directly or indirectly by either the wallboard or the lath under the plaster, so choose the locations for the openings carefully. A location fairly close to joists and studs is best because there the wall materials are strongest.

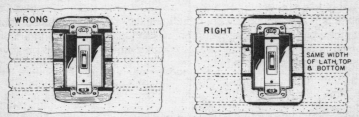

WRONG

RIGHT

SAME WIDTH OF LATH, TOP & BOTTOM

Fig. 16-3. In cutting an opening for a switch box, cut away one whole lath, and part of another lath on each side of the one completely cut. This provides a rigid mounting for the box.

Mounting of Switch Boxes: Let us first discuss the problems if walls and ceilings are of lath-and-plaster construction. In sawing openings for switch boxes, remember that the length of a switch box is approximately the same as two widths of lath, plus the space between the laths. If you remove two complete widths of lath, the mounting brackets on the switch boxes will just barely reach the next two laths. As a result the wood screws by which you attach the boxes (1-in. No. 5 screws are commonly used) will come very close to the edges of the laths, and the laths will split, making a very flimsy mounting and an unsatisfactory job. Remove one width of lath completely, and part of another on *each end* of the opening. Fig. 16-3 shows the wrong and the right way and should be self-explanatory.

Cutting the Opening: Make a mark on the wall, approximately where the switch or receptacle is to be located. Bore a small hole through the mark; with a stiff wire probe to make sure there is no obstruction, and that there is sufficient space all around. Enlarge the hole, to locate the center of one lath; this will locate the center of your opening, up-and-down. Then mark the area of your opening, about 2 by 3¼ in. Drill half-inch holes at opposite corners, and also at the center of top and bottom. The *centers* of the holes must be on the lines of the outline, so that not over *half* of each hole will be outside of the rectangle; unless you watch this carefully, part of the holes may later not be covered by your switch or receptacle plate. See Fig. 16-4.

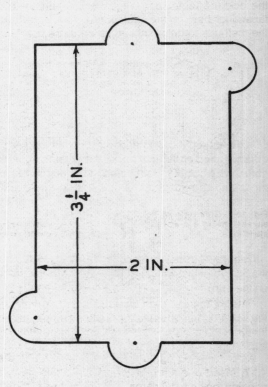

Fig. 16-4. Dimensions of hole for installing a switch box. Make yourself a template the size of the illustration. This is easy to do. Lay a piece of stiff cardboard and a sheet of carbon paper under this page, trace the outline, cut the cardboard to size. This will save you much time.

The holes at the corners are for inserting a hacksaw blade for cutting the opening; the holes at top and bottom are for clearance for the screws for mounting switches or receptacles.

In sawing the opening, use a hacksaw blade, wrapped with tape at one end for a handle. Insert the blade so that the sawing is done as you pull the blade *out* of the wall. If you saw as you push the blade into the wall, you will probably loosen the laths from the plaster, leading to a very flimsy mounting of the box. Hold your hand against the plaster as you pull the blade out.

Installing Switch Boxes in Drywall Construction: Wallboard is not sturdy enough to accept screws as in lath-and-plaster construction, so other methods must be used. In each case adjust the brackets on the ends of the box so that when finally installed, it will be flush with the wall surface. Boxes with beveled corners and cable clamps are preferable but not essential. Bring the cable into the box, tighten the clamps, let about 10 in. project out of the box. If using boxes without clamps, install the connector on the cable, let it project into the box through a knockout near the back of the box, and install the locknut after the box is installed. Be sure the box is deep enough and the knockout far enough back for the cable connector to clear the inner surface of the wall. (Of course the methods to be described can also be used in lath-and-plaster construction.)

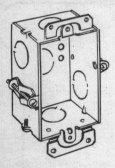

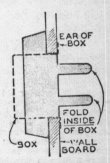

Fig. 16-5. This type of box saves time in old-work installations.

Fig. 16-6. Another way of supporting switch boxes.

Fig. 16-7. A pair of these straps supports a switch box.

One way is to use the special box of Fig. 16-5. It has special clamps on the outsides of the box. After installing the cable, push the box into the opening, then tighten the screws on the external clamps. This makes the clamps collapse, anchoring the box in the wall.

Another way is to use an ordinary box plus the U-shaped clamp of Fig. 16-6. Install the U-clamp with the screw holding it in place, unscrewed about as far as it will go. Slip the box into its opening; the ends of the clamp will expand outwards, and when you tighten the screw holding the clamp, it will anchor the box ears firmly against the wall.

A third method is to use an ordinary box, plus a pair of special straps shown in Fig. 16-7. Insert one strap on each side of the wall opening, and push the box into the opening, being most careful not to lose one of the straps inside the wall. Then bend the short ends of each strap down into the inside of the box. *Be sure* they are bent back sharply over the edge of the box, and lie *tightly* against the inside walls of the box, so that they cannot touch the terminals of a switch or receptacle installed in the box, which would lead to grounds or short circuits.

Mounting Outlet Boxes: If there is open space above the ceiling on which the box is to be installed, and if there is no floor above (or a floor in which a board can easily be lifted as will be explained later) proceed as in new work using a hanger and the usual 1½-in. deep box. The only difference is that you will be working from above.

If all the work must be done from below, the method depends on the ceiling construction, the location of the outlet, and the weight of the fixture. Boxes supporting fixtures weighing more than a few pounds *must* be fastened to the building structure. One method is shown in Fig. 16-8. For light weight fixtures the ceiling itself can support the box, using one of these methods: (1) half-inch deep box, surface mounted, supported by fixture stud on bar hanger (Fig. 9-13) which has been poked up through a hole (which the box covers) and laid *across* the wood laths. See Fig. 16-9. (2) half-inch deep box, surface mounted, supported by toggle bolts through ceiling, either lath and plaster or gypsum board. See Fig. 16-10. (3) 1½ in. deep box with ears, flush mounted, supported by U-clamp. See Fig. 16-11. When using a surface-mounted ½ in. deep box, be sure to select a fixture having a canopy which will cover the box, as shown in Fig. 16-8. In all these cases, fish the cable to the outlet location and fasten it to the box before securing the box.

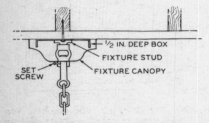

Fig. 16-8. Surface-mounted ½ in. deep box fastened to ceiling joist.

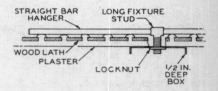

Fig. 16-9. Surface-mounted ½ in. deep box supported from bar hanger by long fixture stud.

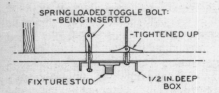

Fig. 16-10. Surface-mounted ½ in. deep box supported from gypsum board by toggle bolts.

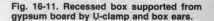

Fig. 16-11. Recessed box supported from gypsum board by U-clamp and box ears.

In all cases, remember that if the ceiling is combustible (can burn), you must cover the space between the edge of the box, and the edge of the fixture, with noncombustible material.

Lifting Floor Boards: Often a board in the upstairs floor must be lifted, to get at the ceiling space. If the flooring is rough, as in ordinary attics, this is no problem. If the lumber is tongued and grooved, it is not so simply done without marring the floor. The first step is to cut the tongue off the boards. A putty knife cut off short so that the blade is only about an inch long, makes an excellent chisel for the purpose. Sharpen the blade and you will have a chisel about 1 or 1½ in. wide, very thin, but short and stubby, which makes it strong. With this you can get down into the crack between two boards and chisel off the tongue as far as necessary. Then bore two holes in the board as close as possible to joists — see Fig. 16-12. With a keyhole saw, cut across just as close to the joist as you can. It is best to cut at an angle so that the board, when it is replaced, more or less forms a wedge. The board should be removed over the space of at least three joists, so that the board when replaced rests directly on at least one joist. When replacing the board, first nail a cleat to the joist at each point where you sawed across. These cleats must be very solidly nailed so that when the board is replaced there will be no springiness.

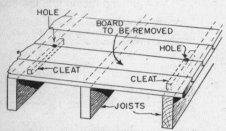

Fig. 16-12. Sometimes floor boards must be lifted. If the boards are tongued and grooved, cut off tongue on each side with very thin chisel, then saw across next to joists.

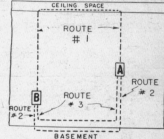
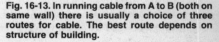

Fig. 16-13. In running cable from A to B (both on same wall) there is usually a choice of three routes for cable. The best route depends on structure of building.

Problem: Two Openings on Same Wall: In Fig. 16-13 cable must run from the opening A to opening B, both in the same wall. Depending on the structure of the building, the cable may run in one of three ways. The simplest way is marked "Route 1" in Fig. 16-13. Use this route if the floor boards can be easily lifted in the floor above, so that the cable can be dropped down from above, to the location of opening A and opening B.

If it is very difficult to get into the ceiling space from above, it may be possible to run the cable down through the basement as shown by "Route 2" in the same picture. If the wall is an outside wall, there will probably be an obstruction where the floor joins the wall. In most cases it is possible to bore upward through this at an angle from the basement. Then, push two pieces of fish tape upward through the bored holes until the ends emerge at A and B. Then, by pulling at A and B, fish the opposite ends of a piece of cable upward until the ends come out at A and B.

If the wall is an inside wall, there may be no partitions in the basement immediately below this wall, so that it should be possible to bore straight upward. Then fish the cable upward to opening A and opening B.

Cable Behind Baseboard: In the problem above, if it is impossible to run the cable through either the ceiling space above or the basement below, use "Route 3" — run the cable behind the baseboard along the bottom of the wall. First remove the baseboard. Then make a small opening into the wall behind the baseboard, directly under *A*, and another directly under *B*. The next step is to cut a channel in the wall between these two openings, forming a trough into which the cable can be laid. Fig. 16-14 shows the completed installation and the details of this picture should make the method clear. Wherever the cable crosses a stud, cover it with a $\frac{1}{16}$ in. steel plate to protect it from future penetration by nails. Finally, replace the baseboard.

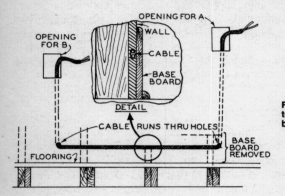

Fig. 16-14. Many times cable can be run to advantage in a trough cut in the wall, behind the baseboard.

Problem: Cable from Opening in Ceiling to Opening in Wall: Our problem in this case is to run cable from an opening in the ceiling which, in Fig. 16-15, has been labeled *C*, around the corner at *D*, and down through the wall to opening *E*. It may be a very simple problem, or it may be a difficult one, all depending upon the construction of the house. If the ceiling joists run in the direction shown in the small inset of the same picture, the problem is greatly simplified. If the floor above is easily lifted, it is then a simple matter to pull the cable in at opening *C*, drop it down at *D* until it comes out at *E*. Even if there is an obstruction at *D*, as is usually the case, it is easy to bore a hole down from above after the board has been lifted.

If it is impossible to lift the floor above, the cable must be gotten around the corner at *D* some other way. An opening must be made into the wall. Sometimes it is made at point marked "No. 1" through the wall, next chiseling away part of the obstruction. Push a length of fish tape into this opening until the end shows up at *C*. Pull it out of *C* until the opposite end is at opening *D*. Then carefully push it down inside the wall until the end shows up at *E*. You then have a continuous fish tape from *C* around *D* to *E*. Attach the cable to the fish tape at one end, pull at the opposite end, and fish it through the wall until you have a continuous cable from *C* to *E*. Sometimes it is easier to do this from the opposite side of the wall, as at point marked "No. 2," boring upward through the obstruction as shown by the dotted arrow. Then using fish tape, pull in the cable as before. After the cable has been fished, patch the wall and the job is finished.

If the opening *E* is not directly below point *D*, but is to the right or left, run the cable over (if the floor board above can be removed) to the proper point above *E* and drop down. If the flooring cannot be removed, drop from *D* down to the baseboard; behind the baseboard run over to a point below opening *E*, and then run upward to *E*.

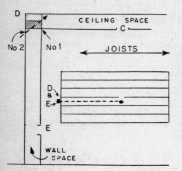

Fig. 16-15. At left, problem in running cable from opening C in ceiling, around corner at D, to E in side wall.

Fig. 16-16. At right, choice of routes for cable in Fig. 16-15, if joists run in the direction shown.

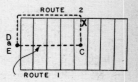

If the joists of the ceiling run in the wrong direction, as shown in Fig. 16-16, there is again a choice of routes. If the floor above can be removed, follow "Route 1," boring holes through the joists through which the cable is to run. If the floor cannot be lifted, make an opening at point *X*, drop the cable down at *X* to the baseboard below, run it behind the baseboard, around the corner to a point below *E*, and from there upward to *E*.

No two houses are alike, so you will simply have to use your own horse-sense in getting around obstructions. Temporary openings often have to be made, and in all cases you will probably use more material than you would for new work. A few extra feet of cable cost much less than the time it takes to follow the shortest route.

Additions to Previous Wiring: Additions to previous wiring are made just as if a complete wiring job were involved. Such small jobs usually consist of adding new outlets or switches. If you have *studied* previous chapters you should be able to make such additions without any trouble, but the following paragraphs will give you some hints for a few specific outlets. *Always turn off the power at the service equipment when working on existing wiring.*

Adding Switch to Existing Outlet: The connections in the present outlet will look a great deal like the left-hand part of Fig. 16-17. There may be more wires in the box than shown, but there will be only two wires connected to the fixture, one of them white, the other black. In the right-hand part of Fig. 16-17 is shown the same outlet after the addition of the switch. To make the proper connections at the fixture, open the black wire splice to the fixture, thus producing two new ends of wire. These two new ends are connected to the two wires in the cable which runs to the switch. The black wire *from the fixture* is connected to the black wire in the new piece of cable; the black wire of the cable which runs up to the original outlet box, is connected to the *white* wire in the *new* piece of cable. This is contrary to general practice, but is the one case where the Code permits a black wire to be attached to a white, and is covered in more detail in connection with Fig. 10-8 in Chap. 10.

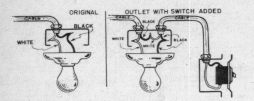

Fig. 16-17. Proper method of adding a single-pole switch to an existing outlet.

If a pair of 3-way switches is to be added, instead of a single-pole switch, proceed as outlined above except add the two 3-way switches as discussed in connection with Fig. 10-14.

Adding Receptacle Outlet: This is very easily done as shown in Fig. 16-18. The end of the cable which in Fig. 16-18 is marked "TO SOURCE," must be run to an existing outlet box which contains a black wire which is *always* hot, and of course also the white wire. If you are in doubt as to whether one of the black wires in any box is always hot, it is a simple matter to check. Turn off the main switch. Remove the cover or the fixture from the box from which you plan to run. Take the tape off the connections and leave the exposed ends of wire sticking out of the box so that they are then accessible, one black wire, one white (of course carefully keeping the bare wires from touching the box). *Turn off the switch which controls that outlet*, and turn on the power at the service equipment.

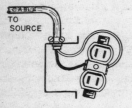  

Fig. 16-18. Wiring a new baseboard receptacle is most simple.

Fig. 16-19. A weatherproof socket makes an excellent test lamp.

Fig. 16-20. Wiring for a ceiling outlet is equally simple.

Then take a test lamp (a lamp in a socket of any type with two short wires attached to it; the weatherproof socket of Fig. 16-19 is very handy for the purpose) and touch these two wires to the wires in the outlet box. Any two wires where the lamp lights (regardless of whether the switch controlling that outlet is on or not), may be used as the SOURCE for the new run of cable. *Be sure to turn the power off again before working on the wires.*

Adding New Ceiling Outlets: A new ceiling outlet may be installed as shown in Fig. 16-20, the two wires marked "TO SOURCE" being run to the nearest outlet with a permanently hot wire, as outlined in the preceding paragraph.

More Outlets: If exposed surface wiring is acceptable, you can use the handy-boxes shown in Fig. 9-5, or special devices that will be shown in Fig. 17-4, using nonmetallic sheathed cable. Follow the instructions in Chap. 17.

Extension Rings: Where the new wiring may be permanently exposed as in basements, it is sometimes convenient to use an extension ring which is like an outlet box without a back. Remove the fixture of the existing outlet, mount the extension ring to the flush box, then run cable or conduit for the new run from the extension ring. Replace the fixture on top of the ring. All this should be clear from Fig. 16-21.

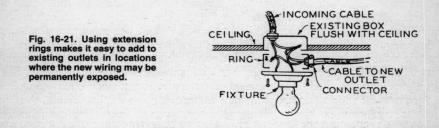

Fig. 16-21. Using extension rings makes it easy to add to existing outlets in locations where the new wiring may be permanently exposed.

Modernizing an Installation: The wiring installed in a house many years ago (and often even if installed only 10 years ago) just is not adequate for the job it is called upon to do today. A complete rewiring job is in order — or is it? Do not jump to the conclusion that every outlet must be torn out, every receptacle replaced. Many times a less expensive job will serve the purpose.

Analyze the Problem: Is the wiring inadequate because you are using too many lights? too many floor lamps? too many radios and TV? That is seldom the case. The wiring usually is inadequate because you have added too many electrical *appliances* that were not allowed for at the time of the original wiring job, probably some that were not even on the market at that time. The installation does not provide enough circuits to operate a wide assortment of ordinary kitchen appliances, plus range, water heater, clothes dryer, room air conditioners. Some of these appliances operate on 240-volt circuits, which may not be available; others operate at 120 volts but when plugged into existing circuits, overload those circuits.

Moreover, the service entrance equipment is just too small for the load, just as 2-lane highways built years ago are too small for the number of cars they are called upon to carry today.

To analyze the problem of your particular house, ask whether *if you disconnected all the appliances*, you have all the *lighting* circuits that you need. The answer is probably "yes" and means that your rewiring job is simplified. You will still have to rewire the house, but probably not as completely as at first appeared necessary. Proceed more or less as if you were starting with a house that had never been wired, but leave the existing *lighting* circuits intact. (These lighting circuits of course will include many receptacles used for small loads like vacuum cleaner, radio, TV, and so on, but *not* the receptacles for appliances in kitchen, laundry, etc.).

Install New Service: Install a new service of at least 100-amp. capacity, with a new outdoor meter, beginning with the insulators on the outside of your house, and ending with new service equipment inside the house. Let the equipment contain as a minimum, branch circuit breakers to protect all the *new* circuits that you are going to install, plus a few spares. Be sure to consult your power supplier about how you plan to proceed. Then follow either Method A or Method B, discussed a little later in this chapter.

Install New Circuits: Install the two small-appliance circuits (or one 3-wire circuit as discussed in Chap. 13). Install the special laundry circuit. Install individual circuits for heavy appliances such as range, water heater, clothes dryer, motor on furnace, etc., all as discussed in Chap. 14. Connect each of these circuits to the breakers in the new equipment that you have installed.

When you have done all this, you will have no power on your new circuits but will still have power on the old circuits. Call your power supplier, *have them disconnect all power on the outside of your house.* Get along for a day or two without electric power, while you reconnect the old circuits into your new equipment.

Methods A and B: Using either method, you must disconnect the incoming service wires from the present equipment. Using Method A, you will continue to use your old service equipment panelboard. Using Method B, you will discard your old panelboard entirely. Using either method, you must disconnect the ground wire from the old equipment and remove it; you must install a new ground wire from the new equipment, to the ground. In the discussion that follows, circuit breakers are mentioned but of course fused equipment may be used.

Method A: Your new equipment will contain a main breaker, plus breakers for all the *new* branch circuits you have installed, plus a few spares for future circuits, and one 2-pole 30-amp. or larger breaker, from which you will run wires to the old equipment. Let's assume your present installation is 3-wire 120/240-volt.

Disconnect the incoming service wires from your old equipment. If they are in conduit, remove the conduit. If they are in cable, remove it so it doesn't enter the cabinet at all. Then run 3-wire cable from the old equipment to the new. The two hot wires of the cable run from the 30-amp. or larger breaker mentioned, to the terminals in the old equipment to which the old service wires were connected. The white wire in the cable runs from the grounded busbar in your new equipment, to the old equipment. The white wires in your old equipment will present a problem. If the old equipment was installed comparatively recently, the grounded busbar in it may be bonded to the cabinet either (a) by a special bonding screw; remove it and the busbar will then be insulated from the cabinet, or (b) by a flexible metal strap bonded to the cabinet, the other end connected to the grounding busbar. Disconnect it from the grounding busbar, which will then be insulated from the cabinet (it will be best to cut off the bonding strap completely). Then connect the white wire from the grounded busbar in the new equipment, to the now-insulated busbar in the old equipment.

If your old equipment was installed a good many years ago [before the schemes described in (a) and (b) above were in use], the grounded busbar was probably bonded directly to the cabinet, with no way of insulating it. Remove it if possible, but in any case remove the white wires from it and install them in a new, insulated, grounded busbar which you must purchase and install, being sure it has enough terminals of the right size to accommodate all the white wires.

If your old equipment is 2-wire 120-volt, the procedure is as already outlined, except that your new equipment, instead of containing one 2-pole breaker to protect the old branch circuits, will contain a single-pole breaker for that purpose. Run a 2-wire cable instead of a 3-wire, from the new equipment to the old. Otherwise the procedure is the same.

Having done all this, it completes the wiring so that your old equipment is now connected to the new equipment. If your old equipment contained *main* breakers or fuses, leave them as they are; they are not required but will do no harm. However, if there are fuse clips that seem deteriorated, or terminals that appear to be in poor condition, remove them.

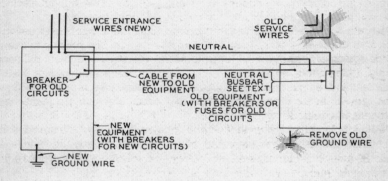

Fig. 16-22. Connections using Method A.

If the original equipment served only lighting circuits (including receptacles), the wires from the old to the new equipment may be No. 10 protected by a 30-amp. breaker. If however the old equipment served and continues to serve also 240-volt loads, install No. 6 wires protected by a 60-amp breaker. All this is shown in Fig. 16-22.

Method B: Remove the original service wires. Disconnect all the branch circuit wires, but don't cut off any of the wires. If the trim or cover of your present equipment has a hinged door in it, or if any openings in it can be closed off, remove and discard the interior (breakers and bus, or fuseholders, terminals, neutral bar, etc.), install knockout seals in any unused openings, and use the enclosure as a junction box. If this is impractical, proceed as follows: If the wiring was using armored or nonmetallic-sheathed cable, remove the locknut inside the cabinet, pull the cable connector out of its knockout, and temporarily screw the locknut on its connector. If the wiring was in conduit, remove the bushing inside the cabinet, pull the conduit out of the knockout, and place the bushing on the end of the conduit. When you have done this on each branch circuit, remove the old equipment completely.

Your new equipment will contain a main breaker, plus other breakers to protect each of the branch circuits, old and new, plus a few spares. But the wires of the old circuits will not reach the new equipment. Where the old equipment used to be located, install a junction box (an empty steel cabinet of a convenient size, 8 by 8 in. or larger as needed, with a steel cover). Then run the wires of the old circuits into this junction box, using the original cables with their connectors, or the original conduits.

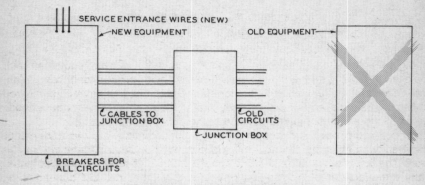

Fig. 16-23. Connections using Method B.

Then run separate cables from the new equipment to the junction box, one for each circuit, and of the same size as the original wires. Connect the new wires to the old, black to black, white to white, using wire nuts or other solderless connectors. See Fig. 16-23. The junction box will not contain a grounding busbar, and the white wires in the box must be carefully insulated from each other, and totally insulated from the box, but the box itself must be grounded.

Call the Power Supplier: When all this has been done, have your power supplier install the meter and connect the power to your new service. Then enjoy the advantages and pleasures of a home wired in modern, adequate fashion.

Chapter 17

FARM WIRING

There is little difference in wiring a house on a farm, as compared with one in the city. But when discussing the problems that come up in wiring the rest of the farm, you must first define what kind of farm you are thinking about. There are many kinds of farms.

The ambitious city worker who chooses to live on a few acres just beyond the city limits, big enough to keep a cow or two and also to raise his own vegetables (with maybe a few left over to sell), will be officially classified as a farmer, despite his full-time city job. Outside of the house, his needs for electric power will be small as compared with those of a full-time farmer raising mostly grain crops. And the grain-farmer's needs will be small indeed as compared with the dairy-farmer's needs.

In this book let us think in terms of farms of some size, and in terms of what they really are: food factories. Let's not think in terms of farm wiring, but rather in terms of wiring for factories. The factories we will be thinking about happen to be food-factories, or farms.

If any manufacturer of clothing or stoves or any other product depended on hand-power to run his machinery, he would go out of business very quickly. To stay in business he depends on the cheapest of all sources of power: electric power. For power in their fields, farmers today depend on tractors. Tractors are not cheap, but nevertheless are better investments than horses. For power in farm buildings, farmers also depend on electric power, but many of them not to the extent that they should, probably because for some farmers electric power is relatively new, and they do not fully understand all the advantages that it has to offer.

Farmers with small electric bills are apt to complain about the expense. Those with large bills do not complain because they know that in buying the power, they are buying one of the biggest bargains available. The more electric power they use, the less other expense they have.

To use large quantities of electrical power requires a well planned and properly installed wiring system and lots of electrical equipment, in total costing in most cases less than a tractor, and still proving to be the better investment. Let's consider some of the things that need discussion in order to properly wire a food-factory, a farm.

Problems in Farm Wiring: The farmer uses about the same amount of power in his house as does his friend in the city. But most farmers use far more power on the rest of their farm, than in the house. There is of course the lighting load of the other buildings, but the real load lies in the other equipment needed in the business of farming. This includes all sorts of motor-driven machinery: water pumps for the entire farm; water heaters, milk coolers, sterilizers and similar equipment for the dairy; extra lighting for forced egg production; incubators and brooders; crop drying equipment; heaters to keep water for livestock or poultry from freezing; and hordes of other equipment.

As a result you must use large wires to carry the heavy amperages without undue voltage drop. Too many overhead wires lead to a messy appearance of the farmyard so you will probably use underground wiring. Heavy loads require heavy service equipment and many circuits. Buildings with livestock present a corrosion and high-humidity problem requiring special attention. These and many other problems will be discussed in separate paragraphs.

Grounds: In city wiring, a single connection to a grounding electrode system (underground metallic water piping, supplemented by an additional electrode) at the house is sufficient if the underground city water system is available for a good ground. (See Chapter 8.) In farm wiring, grounding is done at each building. If the service wires come from a yard pole, one ground must be installed there, another at the house. The Code in Section 250-24 outlines the requirements for other buildings. If another building contains only one circuit, does not contain equipment which must be grounded regardless of circumstances (such as a receptacle), and does not house livestock, no ground is required. It should be noted that farm animals are much less able to withstand shocks than are human beings. Many cattle have been killed by shocks that would be only uncomfortable to a man. For *every* other building, follow one of two methods.

In the first method, install a ground at each building as at the pole and house. This is the only acceptable procedure in any building housing livestock.

In the second method, run an equipment ground*ing* wire (which may be bare, or green if insulated) of the same size as the circuit wires (up to 30 amp. For circuits over 30 amp., or feeders, see NEC Table 250-95). Run it with the circuit wires. This ground*ing* wire of course does not replace the grounded circuit wire. Installing such a ground*ing* wire substantially duplicates an installation using nonmetallic sheathed cable with a ground*ing* wire. The ground*ing* wire must be connected to metallic underground water piping and any other electrodes existing at the building.

If there is underground metal water piping on the farm, use it for the ground if it is near the building, and supplement it with one additional electrode. However, it is not likely that there will be underground piping conveniently located at every point where a ground must be installed, and ground wires must be as short and direct as possible. Therefore often you must use a substitute for the water pipe. There may be existing electrodes, or you may have to provide one. The usual form of farm ground is a special ground rod with a steel core and a copper layer on the outside, driven at least 8 ft. into the ground. It must be at least ½ in. in diameter. Galvanized steel pipe is acceptable according to the Code, but not always to the local inspector. Pipe must be at least ¾ in. trade size, and also driven 8 ft. into the ground. Often two rods or pipes, or longer ones, may have to be driven to secure a low-resistance ground that will serve its purpose. When two rods are used, keep them at least 6 ft. apart, for unless this is done, two rods are little better than one. Naturally they must be connected to each other; use wire of same size as the ground wire.

Do note that if there is less than 10 ft. of metal underground water pipe in direct contact with the earth, it does not qualify as a grounding electrode, but metal water piping inside the building must be bonded to the grounding electrode, so in practice you must bond together the water pipe, regardless of length, and another acceptable electrode (grounded metal building frame, concrete-encased electrode, driven rod or pipe, or buried plate) in every case.

If the ground rod is copper-coated, use a clamp made of copper or brass; if the ground is galvanized iron, use a clamp made of galvanized iron. Iron clamps were shown in Fig. 8-23; one copper type is shown in Fig. 17-1, but there are others similar to those shown in Fig. 8-23.

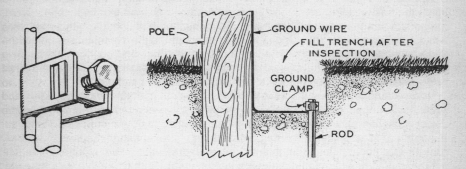

POLE — GROUND WIRE
FILL TRENCH AFTER INSPECTION
GROUND CLAMP
ROD

Fig. 17-1. Use a copper ground clamp with a copper rod.

Fig. 17-2. In some localities, the top of the ground rod is below the surface of the ground. Fill the trench after inspection.

Do use extreme care in installing grounds, so that the installation will not be easily damaged. In some places the custom is to let the top of the ground rod project a few inches from the ground. The ground clamp is permanently exposed. In other localities the ground rod is driven about two feet from the pole (or building) after first digging a trench about a foot deep from rod to pole. The top of the rod is a few inches above the bottom of the trench. The ground wire runs down the side of the pole (or building) to the bottom of the trench, then to the ground clamp on the rod. After inspection the trench is filled in and the rod, the clamp and the bottom end of the ground wire remain buried. See Fig. 17-2. Use the method favored in your locality, but remember that you will have a much better ground if you install two (or three) rods, separated by at least 6 ft., and connected to each other. The ground must be permanent, for the ground connection is a safety device.

Grounds in buildings housing livestock should be installed so that seepage from animal manure does not saturate the ground around the rod. Chemical action in time eats up the wire, the clamp and sometimes even the rod, so that what was once a good ground turns out to be no ground at all.

Use Nonmetallic System: Conduit or armored cable wiring when installed with a permanently good ground constitutes an excellent system with many advantages. On farms, however, poor grounds are the rule rather than the exception, so one of the basic requirements for a good conduit or armored cable installation is missing. Moreover, buildings that house livestock usually have a very *high humidity*, which leads to rusting of all metals, including conduit. The *corrosive action* from the chemicals in the excreta of animals speeds up the process. Experience has shown that when conduit systems are installed in barns and similar locations, the conduit and boxes actually rust away; ground connections sometimes disappear.

Now in a well-grounded and properly maintained conduit installation, an accidentally grounded wire inside conduit or inside a box leads to a blown fuse, indicating trouble. The ground can be located and the fault repaired. But if the conduit is rusted away, near the ground, that leaves you with *ungrounded* lengths of conduit. An accidental ground inside the conduit does *not* lead to a blown fuse; there is no indication of a ground. Then, if a person or an animal touches this ungrounded length of conduit, with a "hot" wire accidentally grounded to the inside of it, what happens? At the minimum, an unpleasant shock; at worst, a dangerous or fatal shock. Very many electrocutions of animals can be traced to this situation. That is what is responsible for the fact that conduit systems are rarely used on farms: nonmetallic sheathed cable is used instead.

Nonmetallic Sheathed Cable in Farm Buildings: Type NM nonmetallic sheathed cable as described in Chap. 10 is a very excellent material when used in *dry locations*. Experience has shown however that when used in moist or corrosive locations, especially in farm buildings housing livestock, its life is very short. Especially where it runs through bored holes in timbers, mildew and rot attack it from the outside. The jute filler cord inside the cable acts like a wick, pulling moisture into the cable from the ends, thus attacking it also from the inside. In due course of time, sometimes after only 2 or 3 years, the cable falls apart leaving a very dangerous, poorly insulated installation, leading to considerable danger of shock and fire. Type NM cable may be used in farm houses, and other buildings that do not have high humidity, but not in buildings where the humidity is normally high, and never in barns or other buildings housing livestock.

For that reason, Type NMC shown in Fig. 10-16 was developed. Each conductor is insulated with its own moisture-resistant insulation; the several conductors are imbedded in a plastic jacket which is specially resistant to moisture, mildew or other fungi, and *corrosion*. Buildings housing livestock have very highly corrosive conditions from the excreta of animals. In such locations Type NMC will outlast Type NM by many years; it will last more or less indefinitely. It may not be buried in the ground, but may be used anywhere that Type NM is acceptable.

In some localities Type NMC is hard to find, in which case use Type UF cable which will be described later in this chapter. Type UF costs more than Type NMC, but may be used anywhere that Types NM and NMC may be used, and may also be buried directly in the ground.

Do use the type with the extra, bare grounding wire that was shown in Fig. 10-16. Chapter 10 also explained how to connect the bare wire. However, if you are using *nonmetallic* boxes described in the next paragraph, the bare wire needs to run only to those boxes that contain receptacles, or which feed a permanently-connected motor or other load, or an item for which grounding is otherwise required. When using nonmetallic boxes, nonmetallic face plates or covers are recommended, but if metal face plates are used, they must be grounded, which is more or less automatic with receptacles (the mounting strap is usually connected to the green, grounding, terminal) but with toggle switches will require a special switch having a grounding terminal on the mounting strap, to which the grounding wire in the nonmetallic sheathed cable must be connected.

Fig. 17-3. Especially on farms, boxes made of an insulating material are frequently used.

Nonmetallic Outlet Boxes: Metal boxes when used in barns and similar locations, tend to rust out just like conduit. Moreover, consider metal boxes used with nonmetallic sheathed cable. Even if the box is in perfect condition, when a hot wire becomes accidentally grounded inside of a poorly grounded box, the entire box becomes "hot." Any person or animal touching the box will receive a shock which can be quite dangerous to humans and even more so to animals. For that reason, boxes of nonmetallic plastic material are being used more and more; their use is a form of insurance. An assortment of them is shown in Fig. 17-3. Use them as you would metal boxes. However, the Code does not require that cable be anchored to single gang switch boxes with connectors where supported within 8 in. of the box measured along the cable sheath, and the sheath extends inside the box at least ¼ in. At other than single-gang nonmetallic boxes, the cable must be secured to the box, as it must to all metal boxes. See Chap. 9.

Likewise popular for farms (as also for surface wiring anywhere) are combination devices shown in Fig. 17-4. Each device is both box and switch, or box and receptacle, and so on, ready to use. Plenty of terminal screws inside each device make it unnecessary to splice wires.

Fig. 17-4. Each of these devices is a combination outlet box plus switch or other device. They are popular in basement, farms and many other locations. Use with nonmetallic sheathed cable.

For the same reason of safety, do not use brass sockets in farm buildings. If you use lights with pull-chain control, be sure there is an insulator in the pull-chain, or at least a piece of cord on the end of the chain. This is a precaution towards safety in case of accidental grounds.

The Meter Pole: On practically all farms today, the power supplier's wires end on a pole in the farm yard. On the pole is found the meter and often a switch or circuit breaker to disconnect the entire installation. The wires are grounded at the pole. From the top of the pole, sets of wires run to the house, to the barn, and to the other buildings to be served. At each building there is a service entrance as described in Chap. 8, without the meter: more about that later in this chapter.

There is a right and a wrong location for the meter pole. Why is there a pole in the first place? Why not run the wires to the house, and from there to the other buildings, as was done when farms were first being wired? That leads to very large wires to carry the total load involved; very large service equipment in the house; expensive wiring to avoid voltage drop which is wasted power; a cluttered farm yard, and many other complications.

Locate the pole as close as practicable to the buildings where the greatest amount of power will be used per year; on modern farms, the house rarely consumes the greatest total. That also means locating the pole so that the largest wires will be the shortest wires. In that way you will find it relatively simple to solve voltage-drop problems without using wires larger than would otherwise be necessary for the number of amperes to be carried. The large expensive wires to the buildings with the big loads will be relatively short, and the smaller less expensive wires to the buildings with the small loads will be relatively long; this keeps total cost down.

From Pole to Building: Wires must run from the pole to individual buildings, and they must be of the proper size. At each building there must be a service entrance quite similar to that discussed in previous chapters, except without a meter. The Code provides a specific method to determine the ampacity of the wire from pole to any farm building with two or more circuits, and rating of the service equipment at that building.

It is assumed that the building will have a 3-wire 120/240-volt service, so the total amperage *at 240 volts* must be determined. For motors, use the amperage shown in the table in Chap. 4. For all other loads, start with the wattage. For lights you can determine the total watts from the size of lamps* you intend to use. For receptacles, if you allow 200 watts for each, you will probably be on the safe side; they will not all be used at the same time. Then divide the total watts by 240, and you will have the amperage at 240 volts.

Caution: Suppose you have in a building, six 120-volt 15-amp. circuits for lights and receptacles. That theoretically makes a total of 6 × 15 or 90 amp. at 120 volts, or 45 amp. at 240 volts. But these circuits will not all be loaded to capacity at the same time, so do not use the theoretical 45 amp. Use the total determined by the preceding paragraph.

For each building to which wires run from the pole (except the farm *house;* figure it as outlined in Chap. 8) first determine the amperage at 240 volts of all loads that have any likelihood of operating *at the same time.* Enter the amperage under *a* of the tabulation below. Then proceed through steps *b, c, d, e* and *f* as outlined in the tabulation.

---

* For fluorescent lights, add about 20% to the wattage, because the watts rating of fluorescents defines only the power consumed by the lamp itself. The ballast adds from 15% to 25%.

a. Amperage at 240 volts of all connected loads that in all likelihood will operate at the same time, including motors if any ...._____amp.

b. If a includes the *largest* motor in the building, add here 25% of the amperage of that motor (if two motors are the same size, consider one of them the largest) ...._____amp.

c. If a does *not* include the largest motor, show here 125% of the amperage of that motor ...._____amp.

d. Total of *a + b + c* ...._____amp.

e. Amperage at 240 volts of all other connected loads in the building ..._____amp.

f. Total of *d + e* ...._____amp.

Now determine the minimum service for the building by one of the following steps:

A. If f is 30 or less, and if there are *not over two* circuits, use a 30-amp. switch and No. 8 wire (No. 8 is Code minimum). If there are *three or more* circuits, use a 60-amp. switch and No. 6 wire.

B. If f is over 30 but under 60, use 60-amp. switch and No. 6 wire.

C. If d is less than 60 *and if f* is over 60, start with f. Add together 100% of the first 60 amp. plus 50% of the next 60 amp. plus 25% of the remainder. For example, if f is 140, add together 60, plus 30 (50% of the next 60 amp.) plus 5 (25% of the remaining 20 amp.) for a total of 60 + 30 + 5 or 95 amp. Use 100-amp. switch and wire with ampacity of 95 amp. or more.

D. If d is *over 60 amp.* start with 100% of d. Then add 50% of the first 60 amp. of e, plus 25% of the remainder of e. For example if d is 75 amp. and e is 100 amp. start with the 75 of d, add 30 (50% of the first 60 of e), and add 10 (25% of the remaining 40 of e) for a total of 75 + 30 + 10 or 115 amp. Use a switch or breaker of not less than 115 amp. rating and wire with corresponding ampacity.

The wire sizes determined above will be the minimum permissible by Code. You would be wise to install larger sizes to allow for future expansion. No. 10 wire is acceptable for spans up to 50 ft., and No. 8 for longer spans. For very long spans, use an extra pole. If the wires are installed in northern areas on a hot summer day, remember that a copper wire 100 ft. long will be a couple of inches shorter next winter when the temperature is below zero. Leave a little slack lest insulators be pulled off buildings during winter.

Wires on Pole: The calculations above will determine the size of the service to each individual building. To determine the size of the wires on the pole, before they break down into smaller sizes to separate buildings, proceed as follows using the amperages determined above:

1. Highest of all the amperages for an individual building, (excluding the house) as determined above, _____ amp. at 100%...._____amp.

2. Second highest amperage, _____ amp. at 75%............._____amp.

3. Third highest amperage, _____ amp. at 65%............._____amp.

4. Total of all other buildings, _____amp. at 50%........._____amp.

5. Total all above ...._____amp.

6. Add the house, as figured in Chap. 8 ...._____amp.

7. Total of 5 + 6 ...._____amp.

Important: If two or more buildings have the same function, consider them as one building for the purpose above. For example if there are two brooder houses requiring 45 and 60 amp. respectively, consider them as a single building requiring 105 amp. If no other building requires more than 105 amp. enter 105 amp. in *1* above.

The total of *7* above is the minimum rating of the switch or breaker (if used) at the pole, and the minimum ampacity of the wires used. Do note that in listing the amperage for any one building, the amperage to use is the *calculated* amperage, not the rating of the switch used. For example if for any building you determined a minimum of 35 amp. but you use a 60-amp. switch (because there is no size between 30 and 60 amp.) use 35 amp., not 60 amp.

Basic Construction at Pole: The three wires from the power line end at the top of the pole. The neutral is usually but not always the top wire; check with your power supplier. The neutral is grounded at the pole as will be explained later. Wires from the top of the pole, down to the meter socket, can be installed in two ways. Use the method favored in your locality.

In one method, called the single-stack construction, the two hot wires and the neutral are run in one conduit down to the meter socket, and two more hot wires are run back to the top of the pole, all five wires in the *same* conduit. The neutral ends at the meter socket, but at the top of the pole it is spliced to all the neutrals of the wires to the various buildings, all as shown in Fig. 17-5.

In the second method, called the double-stack construction, all three wires including the neutral, are brought from the top down to the meter socket in one conduit. Then all three including the neutral run back to the top in a second conduit, as shown in the inset of Fig. 17-5. In this method the incoming neutral of the power supplier's line is *not* spliced to the neutrals of the wires to the buildings; the neutral wire that runs back to the top of the pole from the meter socket of course is so spliced. In this method, two lengths of 3-wire service entrance cable may be used in place of two lengths of conduit.

Both methods are shown in Fig. 17-5. Regardless of the details of the installation, leave at least one-third of the circumference of the pole clear, so that linemen and repairmen can climb the pole without trouble.

Installing the Meter Socket: The meter socket is sometimes furnished by the power supplier, but installed by the contractor. In other localities it is furnished by the contractor. Mount it securely about 5 or 6 ft. above ground level. If a switch or circuit breaker is also used, install it about 5 ft. above ground level, the meter socket just above it. The two "hot" wires from the power line always run to the top terminals in the meter socket. If the usual switch or circuit breaker is used below the meter, the connections are as shown in the inset in Fig. 17-5. If this switch or circuit breaker is not used, the two wires C and D in Fig. 17-5 run directly to the bottom terminals of the meter socket. The neutral wire is connected to the center terminal on the meter, for grounding purposes.

Insulators on Poles: Near the top of the pole install insulator racks of the general type that were shown in Fig. 8-8. Provide one rack for the incoming power wires, one for each set of wires running from the pole to various buildings. Remember that the pull on the wires in a heavy wind, or under ice conditions in northern climates, is terrific. Anchor the racks with heavy lag screws. Better yet, use at least one through-bolt all the way through the pole, for each rack.

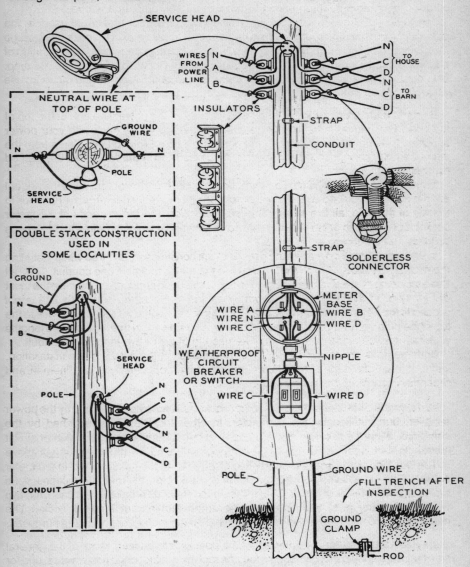

Fig. 17-5. Typical meter-pole installation on a farm.

Installing the Stack: Whether you use a single-stack construction (all wires in one conduit) or the double-stack, the general procedure is the same. At one time the conduit often ended at a point below the insulators, which then required great care in installing the wires to make sure that a drip-loop was provided, as was shown in Fig. 8-19. The purpose of a drip-loop is to prevent water from following the wires into the service head. The Code requirement today is to bring the top of the conduit to a point above the topmost insulator, as shown in Fig. 17-5, thus automatically solving the water problem.

At the top end of the conduit use a service head of the general type shown in Fig. 8-21, with the right number of holes in the insulator. Run wires through the conduit; white for neutral, black or other color (but not green) for the hot wires.

In practice, the switch, meter socket, the conduit with wires inside, and insulators are usually pre-assembled on the pole before the pole is erected. When the pole goes up it is ready for wires to be installed on the insulators.

Connections at Top of Pole: The wires at the top of the pole will be of large size. Splice them using solderless connectors of the types that were shown in Figs. 4-22 and 4-23. Such connectors must be taped, unless you use similar connectors with snap-on insulating covers.

Ground at Pole: At one time it was common practice to run the ground wire out of the bottom of the meter socket, to a ground rod at the pole. Better protection against lightning is obtained if the ground wire runs directly from the neutral at the top of the pole to the ground rod; see Fig. 17-2. Remember, you will have a better ground if you install two rods, separated by 6 ft. or more, and connected to each other. When a ground rod is used (rather than an underground metal piping system), the ground wire is never required to be larger than No. 6. Tuck the ground wire in along the side of the conduit, held by the same straps that hold the conduit in place. In some localities it is stapled to the pole on the side opposite the conduit.

Be sure to protect the ground (and especially the connection of ground wire to ground rod) against mechanical damage. Protect it against damage from livestock, or from vehicles driven too close to the pole. The ground is a safety device; make it a *permanent* ground.

Entrance at House: Install the entrance at the house substantially as was shown in Chap. 8, except that there is no meter to install. Let the service entrance cable or the conduit end at a point above the insulators; from there it runs directly to the service equipment inside the house. Where there is an underground metal water pipe, run the ground wire to it from the neutral bus in the service disconnect enclosure, and supplement the water pipe electrode with an additional electrode, probably a driven rod or pipe, bonding the two together. If there is less than 10 ft. of metal water pipe underground, then the driven rod becomes the primary electrode, but the installation will look much the same, as the water piping (*and* gas piping, if any) *inside* the building must be bonded to the electrode and the neutral.

The wiring inside the house is the same as in other houses, and as already covered in other chapters. Be sure to install plenty of circuits, especially if the farm happens to be one of the small ones in which some of the work of the dairy is done in the house instead of in a special building as on the larger farms.

Entrance at Other Buildings: At each building served by wires from the yard pole, there must be a service entrance similar to that at the house. The important question in every case will be: what size of entrance switch is to be used? That has already been discussed earlier in this chapter. If there are more than two circuits, a 30-amp. switch may not be used, for then the Code requires a 60-amp. switch as a minimum. Be sure to observe Code grounding requirements as outlined earlier in this chapter. Do note that per Code Sec. 230-84(b), Exception, in garages and outbuildings in *residential* property, the disconnecting means may be an ordinary switch (single-pole, 3-way or 4-way) as used in ordinary wiring. On farms, the Code is vague, but presumably the house and other small buildings serving primarily a *residential* purpose, as distinguished from other buildings used primarily in the *business* of farming, would fall into the same category. Buildings serving primarily a residential purpose might include a garage for family passenger cars, a smokehouse or a privy.

All other buildings must have a disconnecting means that will open all the ungrounded wires; it may be located in, on or adjacent to the building. It may be either a switch or a circuit breaker, and *must* be suitable for use as service equipment, as evidenced by the Underwriters' "Service Equipment" label. If the building is served by two 120-volt wires including a grounded wire, use a single-pole switch or breaker; if served by three wires (or only the two hot wires) use a double-pole switch or breaker. How to determine the ampere rating of the switch or breaker has already been discussed.

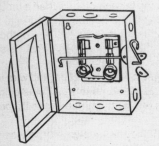

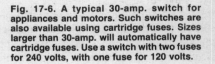

Fig. 17-6. A typical 30-amp. switch for appliances and motors. Such switches are also available using cartridge fuses. Sizes larger than 30-amp. will automatically have cartridge fuses. Use a switch with two fuses for 240 volts, with one fuse for 120 volts.

Branch Circuits in Farm Buildings: How many circuits should a farm building have? Each farm building is part of a business establishment: install enough circuits so that the business of farming can be efficiently carried on. Skimping leads to inefficient work, higher costs in the business of farming. Far better to install a panelboard with space for a few extra circuit breakers, or fuses, than to skimp now and pay a much higher cost per circuit later on.

If the building is to be provided mostly with lights and a few incidental receptacle outlets for miscellaneous purposes, one or two circuits may be sufficient. A machinery shed may require only one circuit; a well equipped farm workshop may require four; a really modern dairy barn with water heaters, milking machines, milk coolers and all the related equipment may require a dozen circuits — and large service-entrance wires in proportion to the load.

The service-entrance switch may then be anything from the simple 30-amp. switch of the general type shown in Fig. 17-6 (if there are not more than two circuits) to a switch of the type that was shown in Fig. 8-5 which provides 12 plug-fuse 120-volt circuits (or a smaller number of 240-volt plug-fuse circuits, each pair of fuses serving one 240-volt circuit), plus a couple of cartridge-fuse circuits for motors, water heaters and so on. The greater the number of circuits that you provide, the less the likelihood of future overloaded circuits, with blown fuses and other troubles. Of course circuit breakers will probably be used instead of fuses.

Clearance above Ground: Be sure to observe the clearances above ground required by the Code in Sec. 225-18: 10 ft. above finished grade, sidewalks or any platform or projection from which wires might be reached; 12 ft. above residential driveways and commercial areas not subject to truck traffic; 18 ft. over commercial or farm properties subject to truck traffic.

Tapping Service Wires at Buildings: If a building contains a quite substantial load, it should be served by wires direct from the pole. But often two buildings are quite close to each other, neither building having a substantial load, and both can then be served by a single set of wires from the pole. Naturally the wires must be big enough for the combined load of both buildings. At the service insulators of the first building, make a tap and run the wires on to the second building, all as shown in Fig. 17-7. At the second building, proceed just as if the wires came directly from the pole.

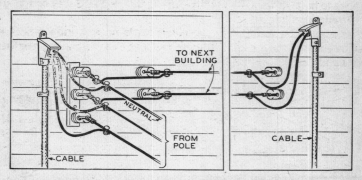

Fig. 17-7. When two buildings are near each other, tap the service wires from pole where they are anchored on the first building. Run them to the second building.

If the second building is very small and requires only 120 volts, tap off only two wires including the grounded wires, as shown in the picture. If the second building has a considerable load so that 120/240 volts are desirable, tap off all three wires. The wires to the next building must be of the same size as the wires from the pole. Remember the requirement for separate service switch and ground at the second building, as discussed in other paragraphs.

Barn Wiring: Barns in general naturally have a great deal of humidity, especially in winter. Proper ventilation will greatly reduce this humidity, but few barns are sufficiently ventilated. The moisture will naturally collect in the coldest parts of the barn, and in winter that means the outside walls. Therefore avoid running cables on outside walls if possible, for the alternate wetting and drying might damage the wiring. However, if you use Type NMC or Type UF cable with nonmetallic boxes, the likelihood of damage is almost entirely eliminated.

It is best not to run cable along the bottoms of joists or other timbers, because it would be more or less subject to mechanical injury. Don't run it at right angles across the bottoms of joists even if running boards are used. The cable will receive far more protection if you run it along the side of a beam, then along the side of a joist to the middle of the aisle to each point where a light is to be installed, as shown in Fig. 17-8. It will take less cable (and lead to less voltage drop) if you run cable down the middle of the aisle through bored holes in the joists; such bored holes should not be near the extreme edge of the joist. The exact method will depend on the details of carpentry in the barn being wired. The important object is to install cable so that is cannot easily be damaged, and to install it where it will be kept away from excessive moisture as much as possible.

Do not skimp on the number of lighting outlets to be installed. The preferable number is one behind each stall; the minimum is one behind each pair of stalls. Do not install light outlets on the bottom of timbers, but preferably between joists so that the bottom of the lamp is flush with the bottom of the joist. Damage to the lamp is less likely to occur that way. Lights should be controlled by toggle switches. It will be found to be a great convenience to have at least some of the lights controlled by 3-way or 4-way switches located at various entrances.

Switches should be installed in protected spots so that they cannot easily be damaged by animals. Mount them at elbow height so that you can operate them even if both your hands are full.

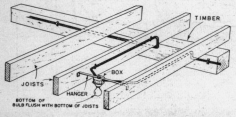

Fig. 17-8. Run cable along a substantial timber to prevent later damage. The bottom of the lamp should not project beyond the bottom of the timber.

Fig. 17-9. Use protected fixtures in hay-mow wiring.

Receptacle outlets should be installed where they will not readily be bumped by animals. Most barns have too few outlets. The right number will depend largely on the kind of barn being wired. Install enough of them so that extension cords need not be used.

Hay-mow Wiring: Be sure to provide a light where it will actually light the stairway or ladder to the hay-mow; this will tend to prevent accidents. Hay-mow lights should be controlled by a switch on the main floor. A pilot light at the switch will be found to be a great convenience.

The dust in a hay-mow is combustible, and a fire can occur if an exposed lamp is broken. When a lamp breaks it burns out, but during that short fraction of a second while it is burning out, the filament is at an exceedingly high temperature (4,000°F) and can start a fire. Therefore for hay-mows, it is best to use gasketed enclosures known as "vaporproof" fixtures, as shown in Fig. 17-9. Glass globes enclose the lamps. In many localities their use, or "dust-ignition-proof" fixtures, is required.

The Code requires that nonmetallic cable must be given special protection when it runs through a floor. For hay-mow wiring, most inspectors will require more protection than the Code calls for. In a hay-mow, there is always danger that a pitchfork will puncture unprotected cable and cause a short circuit. Therefore it is quite reasonable that inspectors require extra protection, from the floor up to a point where hay will never be in contact with the cable. Occasionally an inspector will permit cable installed in the corner formed by wall and stud, with a piece of board nailed over it for protection.

More often the cable is run through a piece of pipe or conduit. Note that this does not constitute "conduit wiring"; the cable is pulled through the conduit and continues to be cable wiring; the conduit merely furnishes mechanical protection for the cable. But it is necessary to ream the ends of the conduit to remove any burrs that might have been formed while cutting it.

In some localities it is the custom to wire the entire hay-mow with rigid or thinwall conduit; the cable then ends on the main floor. This is a safety measure protecting the wiring in the hay-mow against mechanical injury, which might easily occur if cable is installed exposed to hay forks. The conduit is not at all likely to rust out, for in the hay-mow it is not exposed to fumes and moisture as on the main floor. When wiring with conduit, the change-over is easily made as shown in Fig. 17-10.

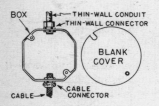

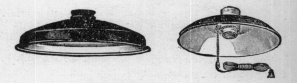

Fig. 17-10. Method of changing from cable to conduit wiring.

Fig. 17-11. Reflectors pay dividends. A 60-watt lamp with a reflector gives as much useful light as a 100-watt without a reflector. Keep reflectors clean.

Code Sec. 250-33 requires that you ground isolated sections of conduit in most cases — and in all cases if over 25 ft. long. That is easily done. Connect the bare grounding wire of the cable to a screw in one of the holes in the outlet box where the conduit begins.

Reflectors: When exposed lamps are used, half of the light goes downward, the other half goes upward and strikes the ceiling. Ceilings in barns and similar buildings are usually dirty, so that most of the light is absorbed, not reflected. As a result nearly half of the light is wasted. Use a good reflector for each lamp, and the half of the light which is otherwise wasted will strike the reflector and be thrown downward. *A 60-watt lamp with a good clean reflector usually gives as much useful light as a 100-watt without the reflector.* Reflectors are inexpensive and some available types are shown in Fig. 17-11. Clean the reflectors regularly.

Poultry Houses: It is well established that if light is provided to lengthen the day in poultry houses, egg production will greatly increase. Install two circuits: One for bright, one for dim light. For the bright, use one 40- or 50-watt lamp with a reflector for every 200 sq. ft. of floor area. For the dim, use 10-watt lamps. Both *must* light up the roosts.

When the time comes to turn the lights off in the evening, first turn off the bright lights; half an hour later turn off the dim lights. Unless you do this the hens will not go to roost but will stay where they are when the lights are turned off. In the morning, turn on the bright lights.

All this of course can be done by hand by the use of proper switches, but it is much better to install a special time switch designed for the purpose. At the time set, the switch turns the lights on in the evening, later dims them, then turns them off. In the morning it turns on the bright lights. An inexpensive time switch is shown in Fig. 17-12.

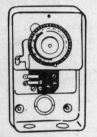

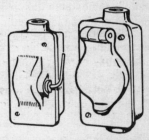

Fig. 17-12. A time switch for poultry house.

Fig. 17-13. Ordinary receptacles and switches may be used outdoors if they are mounted in protective weatherproof housings.

Enclosed, Environmentally Controlled, Poultry and Livestock Confinement Buildings: Accumulation of dust, and particularly dust with moisture, inside ordinary boxes has been identified as the cause of many fires in poultry, cattle, and hog buildings where the animals are continuously housed. In recognition of this problem, new Article 547, Agricultural Buildings, appeared for the first time in the 1978 Code. It is recommended that your local electrical inspector be consulted when there is any question as to whether a particular building is subject to these requirements. The Code says that where animals are permanently housed and there is present: dust from litter or feed; water from cleaning operations or condensation; corrosive vapors from animal excrement; or any combination of these, special precautions must be taken.

Use cast boxes having threaded hubs and close-fitting gasketed covers, similar to those shown in Fig. 17-13, or molded plastic boxes which are similar in appearance. Type NMC or UF Cables must be supported within 8 in. of boxes, and must have water- and dust-tight entries. Use a threaded compression water-tight connector similar to that shown in Fig. 17-14, making sure the opening in the neoprene grommet is the right size for your cable. Rigid metal conduit, intermediate metal conduit, or EMT *may* be used as the wiring method where only dust and/or moisture are present, but are not recommended where corrosive vapors are present.

Motors must be totally enclosed, or so designed as to minimize the entrance of dust, moisture, or corrosive particles. Lighting fixtures must be guarded where exposed to physical damage, must be designed so as to minimize the entrance of dust and moisture, and if exposed to water must be watertight. See Fig. 17-9.

Outdoor Switches and Receptacles: Ordinary receptacles and switches are designed for indoor use. If installed outdoors without further protection, they would be very unsafe. But they can be used outdoors if installed in special waterproof boxes of cast iron or aluminum with threaded hubs instead of knockouts. The left-hand box shown in Fig. 17-13 has only one hub and is called "Type FS"; the one on the right has one hub in each end and is called "Type FSC". Many other hub arrangements are made. Ordinary 15- or 20-amp toggle switches or receptacles up to 50-amp may be installed in them, using the proper weatherproof (gasketed) cover. If using Type UF Cable outdoors be sure to use a weatherproof connector similar to that shown in Fig. 17-14, and also be sure the cable is marked "Sunlight Resistant".

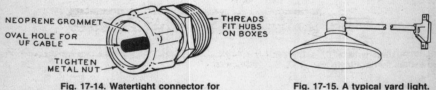

Fig. 17-14. Watertight connector for cable entry into threaded hub.

Fig. 17-15. A typical yard light.

Water Pumps: Every farm worthy of the name will have an electric water system. The many advantages of electric pumps to provide running water need not be repeated here. However, it is well to stress one point: fire protection. A hose connected to a pressure water system is a thousand times as effective as a bucket when it comes to fighting a fire. But in case of fire, it is common practice to pull the main switch, and then the water pump — the fire pump — stops. At the very moment when water pressure is needed as never before, there is no pressure.

The solution is simple. Wire the pump in immediately after the meter, between the meter and the main disconnect switch on the pole. Operating the main switch will kill all electrical wiring on the farm, except the water pump. All this is permissible under Code Sec. 230-72. This kind of installation will require a separate weatherproof switch or circuit breaker for the pump, on the pole, ahead of the main fuses or breaker. Beyond that switch or breaker the wiring is that of an ordinary circuit serving the pump, which has already been covered in Chap. 14.

Yard Lights: Every farm will have at least one yard light; many farms have several. Yard lights are not only useful, but are also a great help in preventing accidents. Fig. 17-15 shows typical construction. The opening at the bottom fits either ½ in. conduit, or a connector for cable or thin-wall conduit.

A yard light should be controllable from at least two points: house and barn. This requires three wires, all as shown in Fig. 13-11. It is contrary to Code to feed a yard light by tapping the wires on the meter pole.

A yard light installed as just outlined requires a considerable quantity of wire and other materials, especially if it is controlled from several points. Ordinary No. 14 wire may be electrically large enough, but often a larger wire must be used for long open overhead wires for mechanical strength. It is of course quite practical to install low-voltage remote-control switching as was discussed in Chap. 13.

Underground Wiring: More and more service entrances, and also wires between buildings, are being put underground, producing neater installations. Underground wiring greatly reduces danger from lightning, and does away with the problem of long spans coming down under ice loads in northern areas.

Underground wiring is done by using conduit, or special cables designed for direct burial. The cables have special moisture-resistant insulation on the conductors, with a very tough outer layer. The Code recognizes two types, available both in single- and multi-conductor.

Fig. 17-16. Underwriters' Type USE cable is designed to be buried directly in the ground without further protection. It is available also as 2- or 3-wire cable.

The most common is Code Type USE (Underground Service Entrance) which has been used for many years. It is shown in Fig. 17-16. Another Type, UF (Underground Feeder), may be used as an underground feeder, but must be *protected by fuses or breaker at the starting point,* therefore it may not be used as service entrance conductors. However, on farms it could be used on underground runs from pole to building, if protected by fuses or breakers.

In the multi-conductor kind, Type UF is substantially the same *in appearance* as Type NMC shown in Fig. 10-16. Nevertheless they are two different kinds of cable. Type NMC may be used in dry, moist, or damp locations, but not for direct burial underground. Type UF may be buried directly in the ground, and may also be used wherever Types NM and NMC may be used.

Do use common sense in installing any kind of underground cable. Code Sec. 300-5 covers the minimum depth to which it must be buried. For *residential* installations the minimum depth is 12 in. provided the cable is protected by a breaker or fuses rated at 30 amp. or less. If not so protected, and in all nonresidential installations, it must be buried a minimum of 24 in. If you use individual wires keep them close together rather than spaced apart. In locations where the cable might be disturbed (as where it crosses roadways, or where it crosses cultivated areas where there may be future digging) bury it deeper and lay a board or similar protective material above the cable before filling the trench. Do not stretch the cable tight; lay it in a sort of "snakey" fashion.

Wherever underground cable emerges from the ground, the above-ground portion must be specially protected against physical damage. Run it through conduit starting about 6 or 8 inches above the bottom of the trench. Install a conduit bushing at the bottom, to protect the cable against damage from possible sharp edges where the conduit was cut. Where the cable enters the conduit, provide a short vertical "S"-curve in each wire, which will tend to prevent damage as the earth settles or moves under the action of frost.

The top may end at the meter socket, in which case the conduit is screwed into the threaded opening in the bottom of the socket. If it ends at a different location *outdoors,* install a service head of the type that was shown in Fig. 17-5. If it ends inside a building, terminate the conduit at a junction box as was shown in Fig. 17-10, and continue with Type NMC cable from that point onward. Of course it may end at the service equipment cabinet. If it does not terminate in a grounded box or cabinet, the conduit must be grounded by a clamp or grounding bushing. If it ends inside a building at a point no higher than the outside end of the conduit, fill the outside end with a sealing compound, to prevent water from flowing into the building.

Lightning Arresters: Lightning causes far more damage to electrical installations on farms, than in other locations. Lightning does not have to strike the wires directly; a stroke near the wires can induce very high voltages in the wires, damaging appliances as well as the wiring. Sometimes the damage is not apparent at once, but shows up later.

Fig. 17-17. A lightning arrester is a good investment on farms.

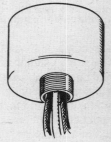

While proper grounding very greatly reduces the likelihood of damage, lightning arresters properly installed reduce the probability of damage to a very low level indeed. Fig. 17-17 shows an inexpensive arrester; it costs not much over ten dollars and is smaller than a baseball. Three leads come out of it; connect the white to the neutral wire, the other two to the hot wires. If the feeders from pole to building are quite long, install another at the building; install it on the service equipment, letting the "neck" of the arrester project into the equipment cabinet through a knockout.

Chapter 18

ELECTRIC MOTORS

A man working hard can deliver no more than about 1/10 horsepower continuously over a period of several hours. Even if the man were paid only $2.00 per hour, it costs at least $20.00 for one horsepower for an hour. At average rates an electric motor will deliver one horsepower for an hour for about six cents. The motor costs little to begin with. It will operate equally well on a hot day or a cold day. It never gets tired, and costs nothing except when running. It uses electric power only in proportion to the power it is called upon to deliver. With reasonable care it will last for many years.

How Motors are Rated: A motor is rated in horsepower. This means that unless it is a special-purpose motor it will deliver the horsepower stamped on its nameplate hour after hour, all day and all week without a stop, if necessary.

Starting Capacity: Motors can deliver far more power while starting than after they are up to full speed. The proportion varies with the type of motor; some types have starting torques 4 or 5 times greater than at full speed. Naturally the amperes consumed during the starting period are much higher than while running at full speed. That means the motor will heat up quickly should it not reach full speed, because of too heavy starting load. Therefore the right kind of motor must be used for each machine, depending on how hard it starts. This will be explained later in this chapter.

Overload Capacity: Almost any good motor after coming to full speed will for short periods develop from 1½ to 2 times its normal horsepower. Thus a 1-hp. motor is usually able to deliver 1½ hp. for perhaps 15 minutes, 2 hp. for a minute, and usually even 3 hp. for a few seconds. No motor should be deliberately overloaded continuously, however, for overloading means overheating; in turn, overheating means short life for the motor. But this ability of a motor to deliver more than its rated horsepower is most convenient. For example, in sawing lumber, ½ hp. may be just right, but when a tough knot is fed to the saw blade, the motor will instantly deliver, if needed, 1½ hp., then drop back to its normal ½ hp. after the knot has been sawed. An automatic water pressure system, designed to operate with about 40 lbs. of pressure in its tank, powered by a ½-hp. motor, may ordinarily require only ½ hp. or less, except for the last few minutes of running while the pressure builds up from 30 to 40 pounds, when ¾ hp. may be needed. The motor will automatically provide the right amount of power.

Gasoline Engines vs. Electric Motors: An engine is usually rated at about 85% of the *maximum* horsepower that it can deliver *while new* at its maximum speed; that speed could be higher than is practical in any given application; moreover, an engine run at its maximum speed would not last as long as one run at more moderate speeds. As it gets older, its maximum horsepower diminishes. Unlike an electric motor, it has no overload capacity. That explains why it is often possible to replace a 5-hp. engine with a 3-hp. electric motor. If the engine always runs smoothly, if it seldom labors and slows down, it can be replaced by an electric motor of a smaller horsepower. On the other hand, if the engine is always laboring at its maximum power, the motor that replaces it should be of the same horsepower as the engine, because no motor should be expected to *continuously* deliver more than its rated horsepower. It won't last long if it does.

Power Consumed by a Motor: The amperage drawn from the power line depends on the horsepower delivered by the motor — whether it is overloaded or underloaded. The watts are not in proportion to the amperes (because in motors, their "power factor" must be considered). You pay for the power in watts, but must provide wire size in proportion to the amperes. As the motor is first turned on it consumes several times its rated current, momentarily. After it comes up to speed, but is permitted to idle, delivering no load, it consumes about half its rated current. Rated current is consumed when delivering its rated horsepower, and more current if it is overloaded.

Speed of Electric Motors: The most common speed for a 60-Hz. motor is a theoretical 1800 rpm. Actually the motor runs at a little over 1750 rpm. while idling, somewhere between 1725 and 1750 rpm. while delivering its rated horsepower. When overloaded the speed drops still more. If overloaded too much, the motor finally stalls. The speed of ordinary ac motors *can not be regulated* by rheostats or switches. Special variable-speed motors are obtainable but they are expensive special-purpose motors and will not be described here.

Reversing Motors: The direction of rotation of a repulsion-induction motor can be changed by shifting the position of the brushes. On other types of ac motors it is changed by reversing two of the wires coming from the inside of the motor. If such a motor must be reversed often, a special reversing switch may be installed for the purpose.

Physical Size and Temperature Rise: A 10-hp. motor made today isn't much bigger than a 3-hp. motor made 40 years ago. This is possible because of advances made in the heat-resisting properties of insulations on the wires used to wind the motor, and the insulations used to separate the windings from the steel in the motor, to reduced air gaps, and to improved magnetic properties of the laminated steel used in the pole pieces. In turn all this means that today's motor will run much hotter than the one made 40 years ago, without being damaged. But always install motors where they will have plenty of air for cooling.

Old-time motors were based on a temperature rise of not over 40°C or 72°F*, over and above the temperature at the motor location, measured before the motor is turned on. Today motors have a "service factor" stamped on their nameplates, ranging from 1.00 to 1.35. Multiply the horsepower by the service factor. The answer tells you what horsepower the motor can safely deliver continuously in a location where the temperature is not over 40°C or 104°F while the motor is not running. That means the motor might develop a temperature of over 212°F, the boiling point of water, but it will not be harmed.

* Do not confuse change in readings of thermometers with their actual readings. When a Celsius thermometer reads 40°, a Fahrenheit thermometer reads 104°. While a Celsius thermometer changes by 40° the Fahrenheit changes by 72°. Thus, if the Celsius changes from 40° to 80°, the Fahrenheit changes from 104° to 176° (by 72°). (What used to be called the centigrade scale is now called the Celsius scale.)

Most motors of 1 hp. or larger have a service factor of 1.15; smaller motors have a higher service factor, some as high as 1.35. But a motor works most efficiently and lasts longer if operated at its rated horsepower.

Dual-Voltage Motors: Larger single-phase motors are usually designed so that they may be operated at either 120 or 240 volts. The motor has four leads. Connected one way, the motor operates at 120 volts; connected the other way, it operates at 240 volts. See Fig. 18-1.

If there is a choice, always operate your motor at the higher voltage. At 240 volts, it will consume only half as many amperes as at 120 volts; with any given wire size, the voltage drop will be only one-quarter as great (when measured as a percentage) on the higher voltage, as on the lower voltage.

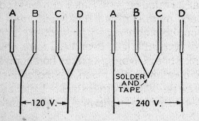

Fig. 18-1. By reconnecting the leads, the motor can be used on either 120 or 240 volts.

Split-Phase Motors: This type of motor operates only on single-phase ac. It is a simple type of motor, which makes it relatively trouble free; there are no brushes, no commutator. It is available only in sizes of ⅓ hp. and smaller. It draws a very heavy amperage while starting. Once up to full speed, the split-phase motor develops just as much power as any other type of motor, but it is not able to start heavy loads. Therefore do not use it to drive any machine which is hard to start, such as a deep-well pump, or an air compressor that has to start against compression. Use it on any machine which is easy to start, or on one where the load is thrown on after the machine is up to full speed. It is entirely suitable for washing machines, grinders, saws and lathes, and general utility use.

Capacitor Motors: This type of motor also operates only on single-phase ac. It is similar to the split-phase type, with the addition of a capacitor or a condenser which enables it to start much harder loads. There are several grades of capacitor type motors available, ranging from the home-workshop type which starts loads from 1½ to 2 times as heavy as the split-phase, to the heavy-duty type which will start almost any type of load whatever. Capacitor motors usually are also more efficient than split-phase, using less watts per horsepower. The amperage consumed *while starting* is usually less than half that of the split-phase type. Capacitor motors are commonly used only in sizes up to 10 hp.

Repulsion-Induction Motors: This type of motor, properly called the "repulsion start, induction run," is commonly called a repulsion-induction or "R-I" motor; it operates only on single-phase ac. It has a very large starting ability and should be used for the heavier jobs; it will "break loose" almost any kind of hard-starting machine. The starting current is the lowest of all the single-phase types of motors.

Three-Phase Motors: These motors are the simplest and most trouble-free type made, and as the name implies, operate only on three-phase ac. Three-phase motors in sizes ½ hp. and larger, cost less than any other type, so by all means use them if you have 3-phase current available. *Do not assume because you have a 3-wire service that you have 3-phase current; more likely you have 3-wire 120/240-volt single-phase current. If in doubt, see your power supplier.*

Universal Motors: This type of motor operates on either dc or single-phase ac. However, it does not run at a constant speed, but varies over an extremely wide range. Idling, a universal motor may run as fast as 15 000 rpm., while under a heavy load the same motor may slow down to 500 rpm. This of course makes the motor totally unsuitable for general purpose work. It is used only when built into a piece of machinery where the load is reasonably constant and definitely predetermined. For example, you will find this motor on your vacuum cleaner, your sewing machine, on some types of fans, on electric drills, etc. The speed of a universal motor can be controlled by a rheostat, as for example on a food mixer or a sewing machine.

Motor Circuits: Every motor must have a disconnecting means, a controller to start and stop it, short-circuit and ground fault protection, and motor overload protection in case of overload or failure to start. Several of these are often combined.

Disconnecting Means: If the motor is portable, the plug on the cord is sufficient. If the motor circuit starts from a circuit breaker in a panelboard, that is sufficient if the circuit breaker is within sight and not over 50 ft. from the controller, *and* the controller is also within sight from and not over 50 ft. from the motor and its driven machinery. It must plainly show whether it is "on" or "off".

If these conditions are not met, install a separate circuit breaker, or a switch of the general type shown in Fig. 17-6. Use a switch with one fuse for a 120-volt motor, with two fuses for a 240-volt motor. If the motor is 2 hp. or less, the switch must be horsepower rated or have an ampere rating at least twice the full-load ampere rating shown on the nameplate of the motor. If the motor is larger than 2 hp. the switch *must* have a horsepower rating not smaller than the horsepower rating of the motor. If a separate disconnecting switch is installed, it must be installed ahead of the controller.

Controllers: A controller is any device to start and stop a motor. On refrigerators, pumps and other equipment with automatically-started motors, it is part of the machine. On manually-started motors it can be a circuit breaker or switch, but is usually what is called a motor starter, as shown in Fig. 18-2. The enclosure for the starter also contains motor *overload devices* which will be discussed later. Controllers for motors over 2 hp. must have a horsepower rating not less than the horsepower rating of the motor.

Use the smaller starter in the illustration for fractional-horsepower motors; it has a manual switch to start and stop the motor. For bigger motors, use the larger starter shown in the same illustration. It has pushbuttons to start and stop the motor. Similar starters have pushbuttons in a separate case; the motor can then be controlled from a distance, wherever the pushbuttons are located.

Motor Overload Devices: It takes many more amperes to start a motor than to keep it running at full speed at its rated horsepower. Also, when a motor is overloaded it consumes more amperes than while delivering its rated horsepower.

A motor will not be damaged by current considerably larger than normal flowing through it *for a short time;* it will burn out if more than normal current flows through it for a *considerable* time. Therefore it is necessary to protect a motor with a device that will permit the high starting current to flow for a *short* time, but will disconnect the motor if current due to overload (or failure to start) flows through it for a *considerable* time. Such devices are called overload devices, and while permitted to be separate, in practice they are usually included in the same enclosure with the starter. Unless a motor is so protected, it may burn out, possibly starting a fire in the process.

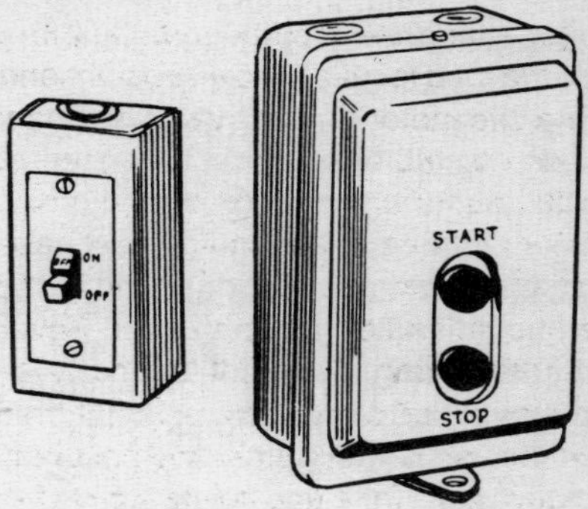

Fig. 18-2. Typical motor starters, with built-in overload devices to prevent damage to the motor in case of overload.

Overload devices include "heaters" and are rated in amperes. When the starter is installed, select overload device heaters for the full-load ampere rating shown on the nameplate of the motor. [In the case of small motors, and often larger motors if they are part of automatically-started equipment such as air-conditioning units, such overload devices are often integral with (built into) the motor.] But whether integral or separate, if the overload device stops the motor, correct the condition that led to the overload. Let the motor and the overload device cool off, then reset manually.

Some overload devices are automatically resetting, but they may not be used where the unexpected restarting of the motor (as for example when powering a table saw) could result in injury to persons.

Motor Branch-circuit Short Circuit and Ground Fault Protection: Motor overload devices, whether built into the motor or installed on the starter, are not capable of interrupting the high amperage that can arise instantaneously in case of a short circuit or ground fault that might occur in the motor circuit, or in the motor. The branch circuit, as well as the controller and the motor, must be protected against such shorts or ground faults by fuses or circuit breakers. The wires in a motor branch circuit must have an ampacity of at least 125% of the full-load motor current, so that they will not be damaged if the motor is overloaded (overload devices do permit up to about 25% overload current for a considerable time, before stopping the motor). The Code (for motors of the kind usually installed in homes and on farms) permits a breaker to have an ampere rating not over 250% of the full-load current of the motor; if time-delay fuses are used, their ampere rating may not exceed 175% of the full-load current.* But, use the smallest rating that will permit the motor to start and operate properly. Fuses that are not of the time-delay type should not be used.

Installing the Motor: Observe the requirement outlined for disconnecting means, controller, and overload units. Use wire with an ampacity of at least 125% of the full-load current of the motor. If the circuit to the motor is rather long, and you use wire that is too small, you may have only 100 volts at the motor during the starting period, instead of more nearly 120 volts. In turn the motor may not start, but if it does it will deliver considerably less than its rated horsepower. So for long circuits, use wire larger than required by Code, to avoid excessive voltage drop.

The table below shows the size of wire to use (assuming it has an ampacity of at least 125% of the full-load current shown on the nameplate of the motor) when the motor is various distances from the service equipment. All distances are one-way. The table is for single-phase motors only.

		ONE-WAY DISTANCE TO MOTOR IN FEET								
Hp.	Volts	No. 14	No. 12	No. 10	No. 8	No. 6	No. 4	No. 2	No. 1/0	No. 3/0
¼		55	90	140	225	360	575	900	1500	2300
⅓		45	75	115	180	300	450	725	1200	1800
½	120	35	55	85	140	220	350	550	850	1400
¾			40	60	100	150	250	400	600	1000
1			35	50	85	130	200	325	525	850
¼		220	360	560	900	1450	2300	3600		
⅓		180	300	460	720	1200	1600	2900		
½		140	220	340	560	875	1400	2200		
¾		100	160	240	400	600	1000	1600	2400	
1	240	85	140	200	340	525	800	1300	2100	
1½		70	110	160	280	400	675	1100	1700	
2		60	90	140	230	350	550	900	1400	2200
3				100	160	250	400	650	1000	1600
5					100	160	250	400	650	1000
7½						110	175	275	450	700

The wires shown in the table are compromises. The voltage drop *while starting* (when the amperage is several times greater than when the motor is running) will be from 3% to 7½%, depending on the type of motor, how hard the load is to start, and other factors. While running, the drop will be about 1½%, less than normally recommended for non-motor loads, but the result of wire sizes large enough to keep the drop within reasonable limits while starting.

* If these values seem to contradict the basic requirement that overcurrent devices may not have an ampere rating larger than the ampacity of the wire being protected, bear in mind that a motor circuit is a special case, where the *overcurrent* device protects only against short-circuits or grounds. Protection against lower values of overcurrent (overload, or failure to start) is provided by the motor over*load* device, as discussed previously.

Problems With Large Motors: The size of farms is constantly increasing, leading to the use of larger sizes of farm machinery, requiring bigger motors: 10 hp., 25 hp. and even larger. In turn this reduces the number of man-hours of labor for a given output, contributing to higher efficiency of labor on the farm, the food-factory, that is continuously being demanded to make the farm profitable.

But most farms have only a *single-phase* 3-wire 120/240-volt service. That requires only two high voltage lines to the farm, and only one transformer. Single-phase motors are not usually available in a size larger than 7½ hp. although a few larger ones are made. But before buying even a 5-hp. single-phase motor, check with your power supplier, to see whether the line and the transformer serving your farm are big enough to operate such a motor.

Single-phase motors 5 hp. and larger require an unusually high number of amperes *while starting,* and the line and transformer often are too small to start such a motor. If you operate the motor only a comparatively few hours per year, your power supplier will object to installing a heavier line and transformer, just as the farmer would not buy a 10-ton truck to haul 10 tons a few times per year, while using it for much smaller loads most of the time.

In a few localities, at least some of the farms are served by a 3-phase line, requiring three wires to the farm, and three transformers. If you are fortunate enough to have 3-phase service, your problems are solved. Simply use 3-phase motors which cost considerably less than single-phase; being simpler in construction, there is rarely a service problem. (Note: If you have 3-phase service there will be available 3-phase power at 240 volts, and also the usual 120/240-volt single-phase for lighting, appliances and other small loads.)

But if you have only the usual single-phase 120/240-volt service, and still need larger motors, what to do? One solution is to use smaller machinery requiring motors not over 3 hp. but that is a step backwards, for it increases the cost of labor, the number of expensive man-hours that must be expended in operating the farm. There is another solution. There are "phase converters" that permit *3-phase* motors to be operated on *single-phase* lines. The phase converter changes the single-phase power into a sort of modified 3-phase power that will operate ordinary 3-phase motors, and at the same time greatly reduces the number of amperes required *while starting.* In other words when operating a 3-phase motor with the help of a phase converter, the same single-phase line and transformer that would barely start a 5-hp. single-phase motor, will start a 7½-hp. or possibly even 10-hp. 3-phase motor; a line and transformer that would handle a 10-hp. single-phase motor (if such a motor could be found), would probably handle a 15-hp. or 20-hp. 3-phase motor.

Phase converters are not cheap, but their cost is partially offset by the lower cost of 3-phase motors, and in any event make possible the operation of larger motors than would be possible without the converter. There are two types of converters: static with no moving parts except relays, and the rotating type. The static type must be matched in size and type with the one particular motor to be used with it; generally there must be one converter for each motor.

The rotating type of converter looks like a motor, but can't be used as a motor. Two 240-volt single-phase wires run into the converter; three 3-phase wires run out of it. Usually several motors can be used at the same time; the total horsepower of all the motors in operation at the same time can be at least double the horsepower rating of the converter. Thus if you buy a converter rated at 15 hp., you can use any number of 3-phase motors totalling not over 30 to 40 hp., but the largest may not be more than 15 hp., the rating of the converter. The converter must be started first, then the motors started, the largest first, then the smaller ones.

But some words of caution are in order. A 3-phase motor of any given horsepower rating, will not *start* as heavy a load when operated from a converter, as it will when operated from a true 3-phase line. For that reason it will often be necessary to use a motor one size larger than is necessary for the *running* load. This will not significantly increase the power required to run the motor, once it is started. The converter must have a horsepower rating at least as large as that of the largest motor.

The voltage delivered by the converter varies with the load on it. If no motor is connected to the converter, the 3-phase voltage supplied by it is very considerably higher than the input voltage of 240 volts. Do not run the converter for significant periods of time, without operating motors at the same time, or it will be damaged by its own high voltage. Do not operate only a small motor from a converter rated at a much higher horsepower, for the high voltage will damage the motor or reduce its life. To be prudent, the total horsepower of all the motors operating at one time should be at least half of the horsepower rating of the converter.

Last but not least, check with your power supplier; some do not favor or permit converters. If they do permit converters, the line and the transformer serving your farm must be big enough to handle all the motors you propose to use.

Chapter 19

HOUSEHOLD "ELECTRICS"

This chapter will be devoted to problems that do not involve the process of actually wiring a building, but rather to problems that arise in a building that was completely wired before you moved into it.

Fuses: How can you tell whether a fuse is blown or not? Look at the top of a new plug fuse. Through the window you will see a small link of metal, usually rather narrow in a small part of its length, as shown in *A* of Fig. 19-1. If the fuse has blown because of an ordinary overload, it will have the appearance of *B* in the same illustration; the narrow part of the link is gone. If the inside of the window is spattered with metal as in *C*, it blew because of a short circuit. In that case you must investigate the cause of the short circuit, before replacing the fuse. The short will probably be found in the cord of a floor or table lamp, or appliance. If in doubt, plug the suspected lamp or appliance into a receptacle protected by a different fuse; if that fuse blows too, you have located your trouble. Repair or replace the lamp or appliance. If, however, one particular fuse keeps on blowing for no apparent reason, the circuit which it protects is probably just simply overloaded. Alternately, the trouble is in the wiring of the circuit which the particular fuse protects, and that short circuit or ground must be located and repaired.

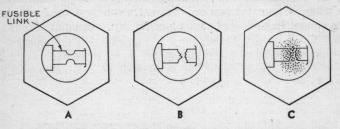

Fig. 19-1. A plug fuse: normal, blown because of overload, and blown because of short circuit.

If your fuses are of the time-delay type as shown in Fig. 5-3, note the coiled spring in a normal fuse. If such a fuse blows, look at the spring: if it is still extended, stretched, the fuse has blown because of short circuit. If the spring is contracted so that its coils touch each other, it blew because of overload.

Circuit Breakers: Most breakers are reset after tripping, as was shown in Fig. 5-6. A few brands, however, are made so that when they trip, the handle returns to the OFF position, and it is only necessary to move it back to the ON position (instead of forcing it *beyond* OFF before returning it to ON, as in most brands).

Label Your Fuses or Breakers: First, mark each fuseholder or breaker with a number. The tools that emboss a letter or a number on adhesive tape are fine for the purpose, but other marking methods will suggest themselves to you.

Then proceed, by trial and error, to determine exactly what outlets are controlled by a particular fuse or breaker. Remove one fuse, or turn off one breaker, and find out exactly which outlets are then dead. When you have done this for all the fuses or breakers, type up a tabulation and paste it securely to the inside cover of your fuse or breaker cabinet. The tabulation might read something like the following.:

1. Main fuses (pullout block).
2. Large bedroom and hall.
3. Small bedroom and bathroom.
4. Receptacle outlets in living room and ceiling lights in kitchen and dining room.
5. Kitchen: bottom halves of all receptacles, and all receptacles in dining room.

6. Kitchen: top halves of all receptacles.
7. All basement lights and receptacles.
8. Range (pullout block).
9. Clothes dryer (pullout block).
10. Water heater (pullout block).
11. Clothes washer.
12. Spare for future new circuit.

This procedure will be found more useful if your home is provided with fuses rather than circuit breakers. When a breaker trips, it is easy to see which one has tripped. When a fuse blows, it is sometimes difficult to determine exactly which one has blown. But if your fuses are "indexed" as suggested, when a light or receptacle is apparently without power, you can go quickly to the right fuse and replace it.

How to Unplug Something: You can yank on the cord and the plug will come out of its receptacle, but *don't do it*. If you do it, you put all the strain on the fine wires inside the cord; sooner or later some of the strands of the wire will break, and you will have a short circuit on your hands. A fuse will blow and your cord will be ruined. Grasp the plug itself and pull it out of the receptacle. In the case of an appliance with a removable cord, it is best not to unplug it at the appliance, but rather at the receptacle. But if you must remove it for example to clean the appliance, again do not pull on the cord; grasp the plug itself to remove it from the appliance.

NEON LIGHT

FOR TESTING CIRCUITS 110V TO 240V AC OR DC

Fig. 19-2. This test light is most useful. It can be used on either 120- or 240-volt circuits.

Using Test Light: While the socket shown in Fig. 16-19, together with a small bulb in it, can be used as outlined in Chap. 16, it is a bit clumsy, and the bulb can easily be broken. There is available a very inexpensive test light, one brand of which is shown in Fig. 19-2. It contains a very tiny neon bulb which consumes only the tiniest fraction of a watt, will last many thousands of hours, and can be used on either 120- or 240-volt circuits.

Is a receptacle live? Just insert the two ends of wire on the tester into the two openings of the receptacle; if the lamp lights, the circuit is live (of course, the receptacle could be worn out, its contacts within the receptacle bent so that they no longer will make good contact with the blades of a plug, and therefore should be replaced). If your receptacles are of the grounding type, as in Fig. 7-11, insert one lead of the tester into the round opening on the receptacle, the other first into one of the two parallel openings, then the other. If the receptacle is properly installed, the tester should light when the second lead of the tester is inserted into the narrow slot that leads to the hot wire to the receptacle, but not when it is inserted into the wide slot which is connected to the grounded wire.

The tester can be used to test fuses in their holders, if you can get at the terminals of the fuseholder. This is especially true of cartridge fuses, which look the same whether blown or not. Touch the two leads of the tester to the two terminals of the fuse holder. If the light in the tester glows, the fuse is blown. If it does not glow, the fuse is not blown. If the cartridge fuses are installed in a fuse-holding pullout block, note that the pullout block usually has small holes through which the leads of the tester can be inserted, without removing the block from its holder.

How to Repair an Extension Cord: The correct answer is: Don't repair it; replace it. An extension cord that is worn, or if the outer jacket is damaged, or if it appears that there is damage within the outer jacket, constitutes a danger of both fire and shock. Using such a cord might be compared to driving a car with thoroughly worn-out tires that may blow out any time; it is an unwise and unsafe policy. All the above applies equally to cords on floor or table lamps, or appliances.

Of course, if a cord is unusable only because the plug on its end has been damaged, but otherwise appears sound, it can be repaired by installing a new plug, if the shortened cord is still adequate to its purpose.

Toasters Can be Dangerous: In most toasters used today the toast is supposed to pop out of toaster when it is properly done. Usually it does. Sometimes it sticks, does not pop out, and that is when the toaster can become dangerous.

Sometimes people just use a fork to release the toast. Don't do it, *without first unplugging the toaster.* The live heating elements are exposed in the toaster, and your fork could touch that heating element. If you should be touching a grounded object (such as a water faucet, or a sink) at the same time, you could receive a dangerous, even fatal shock. Unplug that toaster before reaching inside it for any purpose.

Replacing a Wall Switch: This is a most simple matter. First turn off the main switch or breaker for the building, or at least unscrew the fuse or trip the breaker that protects the circuit on which the switch is installed. *Never work with hot wires.* Remove the face plate over the switch, then remove the two screws that hold the switch to the box. Pull the switch out of the box.

If the switch which you are replacing is a single-pole type, it will have only two terminal screws. Loosen the screws, remove the wires, and connect them to the two screws of your new switch. It makes no difference which wire goes to which terminal. Be sure the loop on your wires is turned in clockwise fashion; see Fig. 4-13. Reinstall the switch in the box and reinstall the face plate.

If the switch you are replacing is the 3-way type, it will have three terminal screws. One of them will be a dark, oxidized color, much darker than the others. In disconnecting the wires, make careful note which wire runs to the dark colored screw of the old switch; it must run to the dark screw on your new switch. It makes no difference to which screws the other two wires are connected.

Some switches do not have terminal screws, but the wires are pushed into openings on the switch. To remove them, just push a screwdriver with a small blade into the slot near the openings for the wires, and you can pull them out. If your new switch has terminal screws, just connect the wires to the terminals as shown in Fig. 4-13.

Replacing a Wall Receptacle: As in the case of replacing a switch, turn off the main switch or breaker of the building, or at least unscrew the fuse or turn off the breaker that protects the circuit on which the receptacle is installed. Remove the face plate, remove the two screws holding the receptacle to the box, and pull the receptacle out of the box. Note that it will have two light, silver-color terminal screws, and two brass terminal screws.

If the receptacle you are replacing is the old-fashioned kind (for 2-prong plugs and without a green terminal screw for a grounding wire) you have two choices: (1) If your house is wired using conduit, or armored cable, replace it with a grounding type receptacle. Run a wire from its green terminal, to the box in which it is installed, unless it is one of the type described in Chap. 7 that is acceptable for use without the grounding wire, or (2) If your house is wired *not* using conduit or armored cable, you must replace the receptacle with another of the nongrounding type.

In either case, note which wire(s) run to the brass terminal screws of the old receptacle, and connect them to the brass screws on the new receptacle. Connect the other wire(s) (which should be white) to the light-colored screws, fold the wires behind the receptacle (See Chap. 15) and reinstall in the box.

If your receptacle is the grounding type for three-prong plugs, be sure the same wire that connects to the green screw of your old receptacle is connected to the green screw of your new receptacle. *Unless this is done, the receptacle wrongly connected can be very dangerous.* Then reconnect the remaining wires: the ones that originally went to the brass screws, to the brass screws of your new receptacle, and the others to the light-colored terminals.

Like switches, some receptacles have no terminal screws. Remove the wires as described under switches. If your new receptacle has terminal screws, just connect the white wires to the whitish screws, the others to the brass screws.

Aluminum Wire: If the 15- and 20-amp branch circuits in your dwelling were wired with aluminum conductors, and: the receptacles and switches are marked "AL-CU", or have no conductor marking, or have push-in or back-wired terminals; or if examination reveals overheating or loose connections, there are some simple steps you should take. Either, 1. replace all switches and receptacles with ones marked "CO/ALR" (see additional discussion on Page 28), using the termination methods discussed in Chapter 4, or 2. "pigtail" a short copper jumper to connect to each switch or receptacle terminal, *being sure to use splicing devices UL listed for use with both aluminum and copper in combination.* This way the aluminum circuit wire will not depend upon the device terminals for continuity. A pigtail splice is shown in Fig. 4-15. Because of reports of field failures at receptacle outlets wired with aluminum, and even though poor workmanship appeared to be the major factor in the failures, both the conductor material and device designs were modified, and presently available wire and CO/ALR devices have been performing well. These changes were completed by November, 1971, so installations made since then should have used the newer materials.

Replacing Lamps in Fixtures: Be sure to observe any maximum lamp size marking on fixtures, and relamp with the proper size. This is particularly important with recessed fixtures and with pan-type fixtures having lamps close to the ceiling. Many fires have started due to the heat from too large lamps.

My Doorbell Won't Ring: The trouble might be in the bell, the transformer, the wiring, or the push-button. In most cases, the push-button is defective. Remove the mounting screws of the push-button; touch the ends of the two wires to the button together, or "jumper" them with a short piece of wire (remember that the voltage is so low that you can't possibly receive a shock). If the bell rings, replace the button.

If the problem is not in the push-button, check the transformer. It will have two screw terminals delivering the low voltage, 6 to 8 volts, for the bell, and too low to give you a shock. Short-circuit these terminals momentarily with a short piece of wire, and if you see a small spark as the short-circuit is applied or removed, the transformer is not defective. You may have to make this test in the dark for the spark might be too small to see in a well-lighted location. If there is no spark, the transformer is defective; replace it.

If both push-button and transformer are ok, check the bell itself. After disconnecting the wires at the bell, remove it, take it to the transformer location, and connect it to the screw terminals of the transformer with short pieces of wire. If it doesn't ring, it is defective, and needs to be replaced. But if it works when temporarily connected, the problem lies in the wiring, and only careful check of the low-voltage wires will locate the defect in the wiring.

If your problem concerns chimes rather than a bell, the first likelihood is that you replaced a bell with chimes, but did not replace the transformer. A bell will operate at 6 to 8 volts, but chimes require a much higher voltage, 18 volts or more. Install a new transformer. And if you test your chimes by moving them to the transformer location, remember they will produce a sound only as the wires are connected; the sound will not repeat until you remove one wire and again touch it to the terminals of your chimes.

In Case of Shock: Throughout this book, the question of safety has been stressed: wire the building strictly according to the Code, and it will be essentially safe. However, shocks, slight or fatal, can still happen, from the use or misuse of appliances, even if they are not defective.

Many fatal shocks have occurred in bathtubs or showers, where the occupant has allowed an appliance to drop into the tub. Even if *not* defective, that can be fatal. Touching a defective radio or heater or other appliance while in a tub or shower can also be fatal. A child might chew on the cord of a lamp or appliance; if he punctures the insulation of the cord, (or if the cord is defective), while also touching a grounded object such as a radiator or plumbing, a fatal shock can result. The housewife who uses a defective appliance while touching a grounded part of plumbing can easily receive a shock.

Preventing shocks is far more important in the long run, than knowing what to do if you discover the victim of a shock. But if you do discover such a victim, the most important point is to be careful, lest you become a second victim yourself. This book cannot possibly give you all the needed information, but can give you a few hints.

We'll first discuss shocks caused by the ordinary 120/240-volt wiring in a house or a farm building. If you find a victim, and it appears that the shock was caused by an appliance or similar equipment, disconnect that equipment. If it has a *wall* switch, turn it off. If it doesn't, pull the plug out of its receptacle; do not touch the cord for it may be defective, but rather grasp the plug and pull it out. DO NOT touch the appliance itself.

In other cases, where the source of the shock is not apparent, hurry up and open the main switch of the building, or throw the main circuit breaker to OFF, if your building is equipped with breakers. Then turn your attention to the victim. If he is breathing, he will probably recover without further aid. But if he is unconscious and not breathing, telephone the police department, fire department, or your hospital, then administer artificial respiration. How to do this is something you should know in any event. If you don't, learn the procedure from the Red Cross or some similar agency.

If the shock has been caused by a high-voltage line that has been knocked down by storm or accident, the situation is quite different, and many more times dangerous. In such a case, phone your power company or police department. If the high voltage wire is touching the victim, *do not* touch him, for if you do, you undoubtedly will be the second victim: you most likely would kill yourself in trying to help somebody who quite possibly is already dead. Theoretically you can use a completely dry wooden pole at least four feet long to push the fallen wire off the victim, but since such accidents seldom occur near your home, you will not have such equipment with you. Depend on experts like power company personnel or police to do the work.

Should you be inside an automobile, over which a high-voltage wire is draped, you will be safer if you stay in the car. If you step out you may place yourself in series with the path of the high-voltage circuit to ground.

INDEX

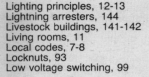

INDEX